Recent Research Trends in Energy Storage Devices

Yogesh Sharma · Ghanshyam Das Varma ·
Amartya Mukhopadhyay ·
Venkataraman Thangadurai
Editors

Recent Research Trends in Energy Storage Devices

Select Papers from IMSED 2018

Springer

Editors
Yogesh Sharma
Department of Physics
Indian Institute of Technology Roorkee
Roorkee, Uttarakhand, India

Ghanshyam Das Varma
Department of Physics
Indian Institute of Technology Roorkee
Roorkee, Uttarakhand, India

Amartya Mukhopadhyay
Department of Metallurgical Engineering and Materials
Indian Institute of Technology Bombay
Powai, Maharashtra, India

Venkataraman Thangadurai
Department of Chemistry
University of Calgary
Calgary, AB, Canada

ISBN 978-981-15-6393-5 ISBN 978-981-15-6394-2 (eBook)
https://doi.org/10.1007/978-981-15-6394-2

This Springer imprint is published by the registered company Springer Nature Singapore Pte Ltd.
The registered company address is: 152 Beach Road, #21-01/04 Gateway East, Singapore 189721, Singapore

Preface

International Meeting on Energy Storage Devices (IMESD)-2018 and Industry-Academia Conclave were convened by Prof. Yogesh Sharma at Indian Institute of Technology Roorkee, Uttarakhand, India, during December 10–12, 2018. IMESD-2018 was a debut event in this international meeting series, which was considered a milestone event in India for communities working in the field of energy storage devices. About 300 papers covering broad areas of solid-state ionics/nanoionics, rechargeable secondary batteries, particularly Li-ion, Na-ion batteries, supercapacitor/flexible supercapacitor/hybrid supercapacitor, nanostructured energy storage materials. synthesis, characterization and applications, electrode/electrolyte interface and associated phenomenon, fast ionic conductors: liquids, polymer, gel and solid electrolytes, modeling and simulation of energy storage devices, safety, reliability, cell design engineering, fuel cell and technologies and hydrogen storage devices were received and presented. This book is a compendium of selected papers presented at the conference and conclave.

Participants from fourteen countries had discussed the issues related to role of energy storage devices to reduce the carbon footprint to save the planet in the line of Paris Agreement. Eminent scientist and founder of many companies, co-inventor of many materials being used in commercialized battery system, Prof. Arumugam Manthiram from University of Texas at Austin, a well-known researcher, authored/co-authored more than 800 research articles emphasized on the need of efficient trained manpower to deal with the development of indigenous battery technology. During a popular lecture by Prof. Vijaya Mohanan K. Pillai, a policymaker, adviser and outstanding scientist (former director of CSIR-CECRI and NCL, Pune) suggested an urgent need for development of affordable material and energy storage technologies to support our electric vehicle program of self-sustainable growth of whole world.

Prof. Ashok Jhunjhunwala (Bhatnagar awardee and Padamshree from IIT Madras, leading researcher/entrepreneur/administrator and advisors to many ministries/organizations and a leading person to make this electric mobility program of Govt. of India successful, emphasized on the need of developing robust charging platform and battery recycling technologies of battery material along with indigenous development of secondary batteries specifically for EV application. During discussion/deliberation at Meeting and Conclave, most of the companies/industries working in the area of storage devices such as Amar Raja Batteries Ltd. TVS Motors, XNRGI (Exponential Energy), USA, BHEL, BEL MNRE, DRDO have shown keen interest to participate in energy storage program/facility at IIT Roorkee.

Roorkee, India

Prof. Yogesh Sharma
Convener, IMESD-2018

Contents

Flexible Fibre Supercapacitor Using Synthesized Biomass-Based Activated Carbon and Few-Layer Graphene for Wearable Electronic Devices 1
Ankit Singh, Kaushik Ghosh, Sushil Kumar, Ashwini K. Agrawal, Manjeet Jassal, Pranab Goswami, and Harsh Chaturvedi

Combustion Synthesized $MLi_2Ti_6O_{14}$ (M = Sr, Ba, Pb) Titanate Anodes for Lithium-Ion Batteries 9
Anshuman Chaupatnaik and Prabeer Barpanda

Role of Annealing Temperature on Structural Modification of MoO_3 for Enhanced Electrochemical Properties 19
Arvind Singh, Bilal Ahmed, Saurav K. Ojha, and Animesh K. Ojha

Reversible Sodium and Potassium-Ion Intercalation in $Na_{0.44}MnO_2$ 27
Baskar Senthilkumar, Sai Pranav Vanam, and Prabeer Barpanda

Cobalt–Phosphate-Based Insertion Material as a Multifunctional Cathode for Rechargeable Hybrid Sodium–Air Batteries 35
Chinnasamy Murugesan, Baskar Senthilkumar, Kriti Choudhary, and Prabeer Barpanda

Investigation of Charging and Discharging Characteristics of Lithium-Ion Batteries for Application in Electric Vehicles 43
Dipanwita Bhattacharjee, Pankaj Kalita, and Sudipta Bijoy Sarmah

Comparative Studies on Crystalline and Amorphous Vinylidene Fluoride Based Fibrous Polymer Electrolytes for Sodium-Ion Batteries 55
S. Janakiraman, Mohammed Khalifa, Rasmita Biswal, Sudipto Ghosh, S. Anandhan, and A. Venimadhav

Free-Standing Bi-Induced ZrO_2 Nanoflake Array Photoanodes Fabrication for Photoelectrochemical (PEC) Water Splitting Applications . . . 65
B. Jansi Rani, A. Anusiya, G. Ravi, and R. Yuvakkumar

Silicon/Acetylene Black–Carbon Composite as an Anode for Lithium-Ion Battery . . . 73
Jay Krishan Dora, Sudipto Ghosh, and Natraj Yedla

Layered $Na_2Mn_3O_7$: A Robust Cathode for Na, K, and Li-Ion Batteries . . . 81
Krishnakanth Sada, Baskar Senthilkumar, Ritambhara Gond, Valerie Pralong, and Prabeer Barpanda

Structural and Electrical Transport Investigations on $Mg(ClO_4)_2$-PEO-PEG Solid Polymer Electrolyte Prepared by Hot Press Assisted Route . . . 89
Munesh Rathore, M. Dinachandra Singh, and Anshuman Dalvi

Effect of Au Decoration on the Performance of Perovskite Solar Cells . . . 97
Nilesh G. Saykar, Anil Arya, Indrani Banerjee, and S. K. Mahapatra

In-situ Fabrication of Alternately Stacked MoS_2/Au Multilayered Thin-Film Electrodes for Electrochemical Energy Storage Application . . . 105
Pranjala Tiwari and Ramesh Chandra

Conductivity and Dielectric Spectroscopy of Na^+ Ion Conducting Blended Solid Polymer Nanocomposites . . . 115
Pritam, Anil Arya, and A. L. Sharma

In Situ X-Ray Diffraction and Alkali Ion (A = Li, Na, K) Intercalation Behavior of $Na_2FeP_2O_7$ Pyrophosphate . . . 125
Ritambhara Gond, Shashwat Singh, Sai Pranav Vanam, Valerie Pralong, and Prabeer Barpanda

Fabrication and Electrochemical Performance of Low-Cost Soluble Lead Redox Flow Battery Using Two Different Carbon Electrodes . . . 133
Shivangi Kosta, R. Sneha, and Kuldeep Rana

Development of Novel Ternary SnSb-ZnO Nanocomposites as Alternative Anode Material for Lithium-Ion Batteries . . . 147
Shivangi Rajput and Amrish K. Panwar

Structural, Electrochemical, Thermal and Electrical Performance of $LiMnPO_4$ Cathode Material for Lithium-Ion Batteries . . . 159
Snigdha Sharma, Amrish K. Panwar, and M. M. Tripathi

About the Editors

Yogesh Sharma is an Associate Professor in the Department of Physics, Indian Institute of Technology (IIT), Roorkee. He pursued his M.Tech. and Ph.D. from IIT Roorkee and the National University of Singapore, respectively. His areas of research interest lie in experimental condensed matter physics with special relevance to energy storage. He has authored more than 45 research articles in reputed journals and is a member of many professional societies, including the Electrochemical Society (USA), Material Research Society Singapore and the Indian Physics Association.

Ghanshyam Das Varma is a Professor in the Department of Physics, Indian Institute of Technology (IIT), Roorkee. He pursued his M.Sc. and Ph.D. from Banaras Hindu University (BHU). He has extensive research experience in the area of experimental condensed matter physics, centring on the synthesis and structural characterization of high temperature superconductors, colossal magnetoresistive materials and carbon nanotubes. He has authored more than 60 research articles in reputed journals.

Amartya Mukhopadhyay is an Associate Professor in the Department of Metallurgical Engineering and Materials Science, IIT Bombay. He has done his M.Tech. and Ph.D. from IIT Kanpur and the University of Oxford, UK, respectively. His areas of research interest include materials for electrochemical energy storage and advanced structural ceramics/composites. He has authored 4 book chapters and more than 70 research papers in reputed journals, as well as holds 3 patents.

Venkataraman Thangadurai is a Professor and Associate Head in the Department of Chemistry, Faculty of Science in the University of Calgary, Canada, and the Associate Director of Calgary Advanced Energy Storage & Conversion Research – Technology. He pursued his Master's and Ph.D. from Muthurangam Government Arts College (University of Madras), India and Indian Institute of Science, Bangalore, India respectively. His research interests are in the area of solid-state ionic technologies, and in particular, materials development for energy conversion, storage and environmental applications. He has been the recipient of the Keith Laidler Award by the Canadian Society for Chemistry and the Outstanding Invention of 2013 by the University of Maryland, USA. He holds 4 patents and has authored more than 130 research papers in reputed journals and 2 book chapters.

Flexible Fibre Supercapacitor Using Synthesized Biomass-Based Activated Carbon and Few-Layer Graphene for Wearable Electronic Devices

Ankit Singh, Kaushik Ghosh, Sushil Kumar, Ashwini K. Agrawal, Manjeet Jassal, Pranab Goswami, and Harsh Chaturvedi

Abstract We have fabricated a flexible fibre supercapacitor having twisted architecture incorporating synthesized carbonaceous electrode materials from widely available inexpensive biomass (banana peel) for energy storage which can be utilized for powering emerging wearable electronic textiles and devices. Activated carbon and few-layer graphene were synthesized by carbonizing KOH impregnated and ethanol/acetone washed banana peels, respectively, at high temperature under inert atmosphere. The synthesized carbonaceous electrode materials along with TiO_2 nanopowder were deposited on conductive carbon fibre followed by dip coating with gel electrolyte and twisting to develop the supercapacitor. The fabricated fibre supercapacitor demonstrated high specific capacitance of 15.45 F/g and volumetric capacitance of 1.77 F/cm^3 at 10 mV/s scan rate. The developed supercapacitor retained ~92% of capacitance upon bending. The fibre supercapacitor can be fabricated on large scale using simple methods and inexpensive biomass for powering wearable electronic applications.

Keywords Flexible · Fibre supercapacitor · Inexpensive biomass

1 Introduction

Recent progress in wearable electronics has affirmed its promising future. The key challenge inhibiting its wider acceptance is lack of wearable energy storage device. Supercapacitor has long life, is safe and environmentally friendly and can prove to be a good alternative for wearable energy storage applications [1]. Fibre supercapacitor exhibit unique advantages as compared to existing solid-state planar supercapacitor

A. Singh · P. Goswami · H. Chaturvedi (✉)
Centre for Energy, Indian Institute of Technology Guwahati, Guwahati, India
e-mail: harshc@iitg.ac.in

K. Ghosh · S. Kumar
Institute of Nano Science and Technology, Mohali, India

A. K. Agrawal · M. Jassal
SMITA Research Lab, Indian Institute of Technology Delhi, Delhi, India

Y. Sharma et al. (eds.), *Recent Research Trends in Energy Storage Devices*,
https://doi.org/10.1007/978-981-15-6394-2_1

in terms of energy density, flexibility, lightweight, varied architecture and applications [2, 3]. We demonstrate fibre supercapacitor having a twisted architecture with asymmetric electrode configuration, using conducting carbon fibre incorporated with TiO_2 nanopowder and synthesized biomass-based activated carbon and few-layer graphene as active electrode materials. Fibre supercapacitor can be woven into the fabric for flexible/wearable energy storage hence eliminating the requirement of carrying bulky devices. This fibre supercapacitor can be utilized in the domain of smart textiles, biomedical and other wearable electronic applications [2, 3].

2 Experimental

Locally available inexpensive source of biomass like banana peels were washed thoroughly with distilled water and dried at 80°C overnight in vacuum oven which was used for synthesizing activated carbon and few-layer graphene.

2.1 Synthesis of Activated Carbon

The dried banana peels were then impregnated with aqueous solution of KOH which was prepared by mixing 3 g of KOH flakes with 50 ml of DI water for 2 h. The impregnated peel was dried at 80°C overnight in vacuum oven. The dried impregnated banana peel was carbonized at 600 °C for 1 h inside a muffle furnace in absence of air. The carbonized residue was washed with dilute HCL and DI water until pH 7 was attained and dried at 80 °C overnight in vacuum oven resulting in activated carbon. The activated carbon was observed using FESEM as shown in Fig. 1 showing its

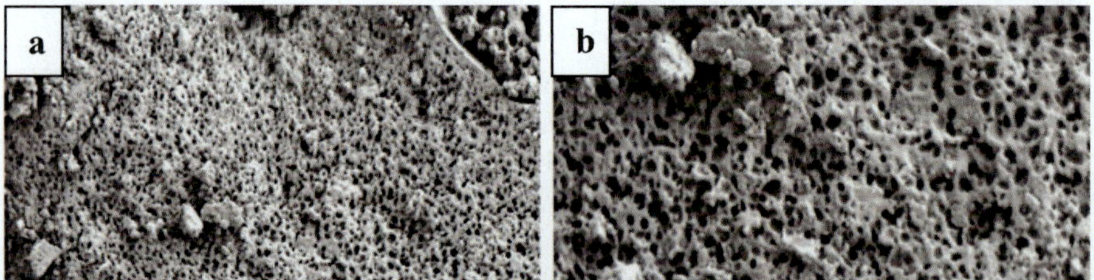

Fig. 1 FESEM image of activated carbon at (**a**) 5kX and (**b**) 10kX magnification showing well-porous structure

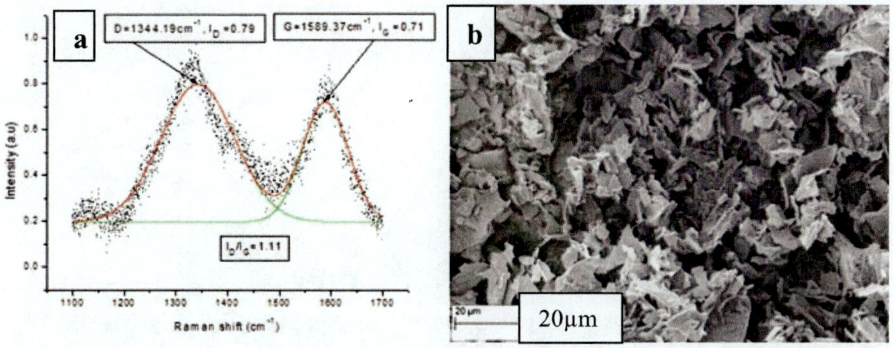

Fig. 2 **a** Raman spectrum of few-layer graphene at 633 nm, **b** FESEM image of few-layer graphene at 2kX magnification

well-porous structure. The synthesized activated carbon also has BET surface area of 62.03 m^2/g [4].

2.2 *Synthesis of Few-Layer Graphene*

The dried banana peels were further washed with ethanol/acetone and left to dry at room temperature. The peels were carbonized at 950 °C in a tubular furnace for 2 h in presence of inert argon gas. The carbonized residue was crushed in mortar-pestle resulting in few-layer graphene which was verified by Raman spectroscopy (Fig. 2a) and FESEM investigation (Fig. 2b) [5].

2.3 *Fabrication of Fibre Supercapacitor*

The schematic structure of fabricated twisted fibre supercapacitor is shown in Fig. 3a. A pair of conductive carbon fibre was painted with conductive silver ink (Fig. 3b) followed by deposition of equal amounts (1 mg of each constituent) of TiO_2/ few-layer graphene and activated carbon/ few-layer graphene (Fig. 3c) forming the cathode and anode, respectively. The Cross section of silver-painted carbon fibres was observed by FESEM at (Fig. 3d) 250X and (Fig. 3e) 1kX magnification showing individual carbon fibre having an approximate diameter of 7 μm. The electrodes were dip coated with the PVA/H_3PO_4 gel electrolyte twice. The gel electrolyte was prepared by mixing 1 g of PVA in 10 ml of DI water at 90°C under constant magnetic stirring followed by addition of 1 ml of H_3PO_4 acid [6]. The electrodes were twisted and dip coated with the gel electrolyte to have better adhesion and dried overnight at room temperature forming the twisted fibre supercapacitor (Fig. 3b).

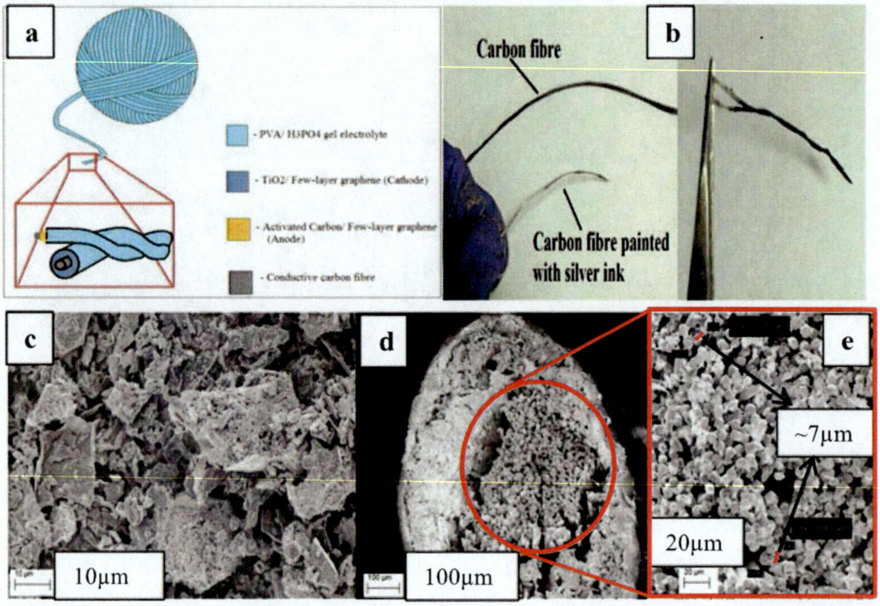

Fig. 3 **a** Schematic representation of twisted fibre supercapacitor, **b** optical image of fibre supercapacitor fabricated using pair of conductive carbon fibre as substrate, FESEM image of **c** activated carbon/few-layer graphene at 3kX, cross section of silver-painted carbon fibres at **d** 250X and **e** 1kX magnification showing individual carbon fibre having an approximate diameter of 7 μm

3 Device Characterization

The fabricated fibre supercapacitor was characterized by cyclic voltammetry (CV) and charge–discharge (CD) tests by using Metrohm Autolab electrochemical workstation. CV tests were carried out at scan rates ranging from 10 to 200 mV/s (Fig. 4a). From the CV profile, specific capacitance and volumetric capacitance (Fig. 4b) were calculated at respective scan rates using formulae,

Specific capacitance (CV curves),

$$\mathrm{Csp} = \frac{A}{m * v * V}$$

Volumetric capacitance (CV curves),

$$\mathrm{Cv} = \frac{A}{\Delta v * v * V}$$

where A is area of CV curve, m is mass of active electrode material (4 mg), Δv is the volume of the device (0.035 cm^3), v is scan rate in mV/s and V is potential window (1 V). The fabricated supercapacitor demonstrated specific capacitance of 15.45

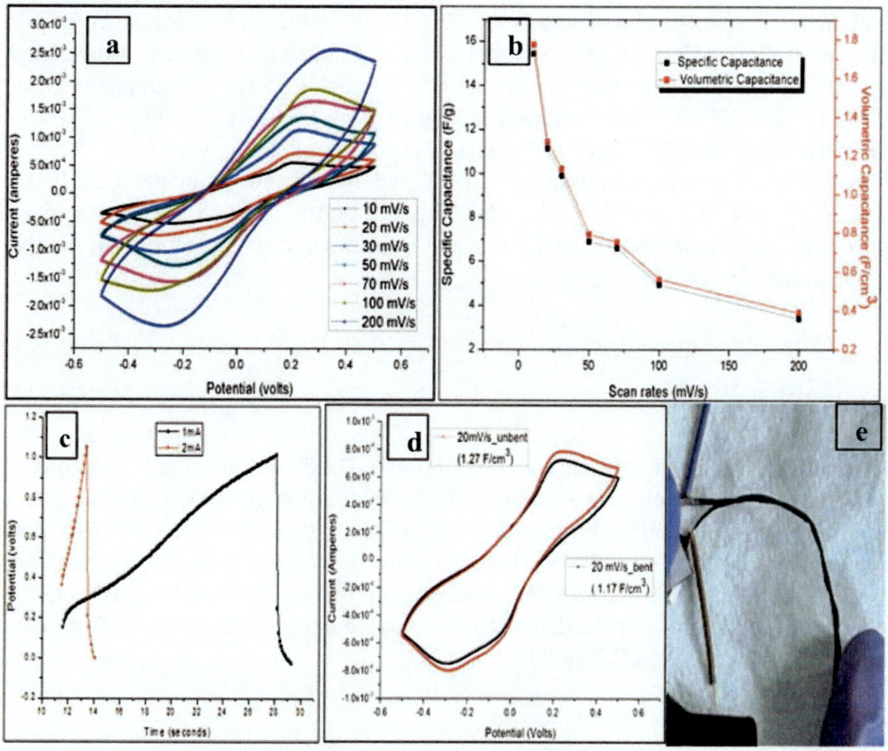

Fig. 4 **a** CV profile at different scan rates, **b** specific capacitance and volumetric capacitance plot and **c** CD plots of the fabricated fibre supercapacitor, **d** supercapacitor retained 92% of capacitance upon bending at 20 mV/s scan rate, **e** fibre supercapacitor under bent conditions

F/g and volumetric capacitance of 1.77 F/cm^3 at 10 mV/s scan rate. The fabricated supercapacitor was also tested for charge–discharge (CD) at 1 and 2 mA discharge currents (Fig. 4c). Specific capacitance and volumetric capacitance were calculated at respective discharge currents using formulae,

Specific capacitance (CD profile),

$$\mathrm{Cs} = \frac{I * \mathrm{td}}{m * V'}$$

Volumetric capacitance (CD profile),

$$\mathrm{Cv} = \frac{I * \mathrm{td}}{\Delta v * V'}$$

where I is the constant discharge current, td is discharge time, m is mass of active electrode material (4 mg), Δv is the volume of the device (0.035 cm^3) and V' is effective discharge voltage (excluding IR drop). The fabricated supercapacitor demonstrated specific capacitance of 1.27 F/g and volumetric capacitance of 145 mF/cm^3 at 2 mA (57 mA/cm^3) discharge current. Capacitances of few of the earlier reported twisted fibre supercapacitors are CNT based (13.5 F/cm^3), graphene based (40 F/g), IR-CNT@PANI electrode (78 F/g) (Tables 1 and 2) [2] and all graphene (GF@3D-G) based (30–40 μF at 2μA), CNT/MnO_2 composite yarn (25.4 F/cm^3 at 10 mV/s), Asymmetric electrodes of Co_3O_4 nanowires and graphene (1.59 F/cm^3 at 40 mA/cm^3) (Table 1) [3]. The fibre supercapacitor retained ~92% of capacitance under bent condition at 20 mV/s scan rate demonstrating its ability for flexible applications (Fig. 4d, e). [7]

4 Materials

Titanium (IV) oxide (TiO_2) nanopowder (718467-100G) was obtained from Sigma-Aldrich and has particle size <21 nm (TEM). Potassium hydroxide (KOH), flakes (2815.20.00–500G), orthophosphoric acid (H_3PO_4), 88% (O11329– 500 ml), poly vinyl alcohol (PVA), 98% (3905.30.00–500G) was obtained from NICE Chemicals Pvt Ltd, Kerala. Silver ink (200 ml) was obtained from SNAB GRAPHIX (India) Pvt Ltd, Bangalore. Conductive carbon fibres were obtained from SMITA research lab, Textile Department, IIT Delhi.

5 Conclusion

Fibre supercapacitor having twisted architecture was fabricated using synthesized activated carbon and few-layer graphene from banana peels along with TiO_2 nanopowder as active electrode materials and conductive carbon fibre as substrate. Porous activated carbon increased the electrode surface area, and the few-layer graphene increased the conductivity which collectively enhanced the energy-storing capability of the fabricated device. The fabricated supercapacitor demonstrated specific capacitance of 15.45 F/g and volumetric capacitance of 1.77 F/cm^3 at 10 mV/s scan rate. The twisted supercapacitor also retained ~92% of capacitance upon bending. Other transition metal oxides like MnO_2, NiO, Co_3O_4 as electrode material should be explored to further increase the energy-storing capacity. Coaxial architecture of fibre supercapacitor will further enhance the capacitance and will have better mechanical stability. The fabricated fibre supercapacitor can be used in the domain of smart textiles, biomedical and other wearable electronic applications.

Acknowledgements We would like to thank Centre for Energy, Central Instruments Facility of Indian Institute Of Technology, Guwahati. We would like to acknowledge Ucchatar Avishkar Yojna (UAY), MHRD, Government of India. We would also like to thank G Labs Pvt ltd, Kolkata, for their support.

References

1. Conway, B.E.: Transition from supercapacitor to battery behaviour in electrochemical energy storage. J. Electrochem. Soc. **138**(6), 1539–1548 (1991)
2. Senthilkumar, S.T., et al.: Advances and prospects of fiber supercapacitors. J. Mater. Chem. A **3**, 20863 (2015)
3. Wang, X. et al.: Flexible fiber energy storage and integrated devices: recent progress and perspectives. Mat. Today **18**(5), 265–272 (2015)
4. Wang, et al.: KOH activation of carbon-based materials for energy storage. J. Mater. Chem. **22**, 23710–23725 (2012)
5. Purkait, T. et al.: Large area few-layer graphene with scalable preparation from waste biomass for high-performance supercapacitor. Sci. Rep. **7**, 15239 (2017)
6. Chen, et al.: Effect of different gel electrolytes on graphene-based solid-state supercapacitors. RSC Adv. **4**(68), 36253–36256 (2014)
7. Boruah, B. et al.: Flexible array of microsupercapacitor for additive energy storage performance over a large area. ACS Appl. Mater. Interfaces **10**, 15864−15872 (2018)

Combustion Synthesized $MLi_2Ti_6O_{14}$ (M = Sr, Ba, Pb) Titanate Anodes for Lithium-Ion Batteries

Anshuman Chaupatnaik and Prabeer Barpanda

Abstract Three isostructural $MLi_2Ti_6O_{14}$ (M = Sr, Ba and Pb) titanates with an orthorhombic structure have been considered as anodes for lithium-ion batteries. Phase pure white titanate powders were prepared using a rapid two-step solution combustion method. Morphology and electrochemical activity of the titanate powders have been comparatively studied. 3 Li+ could be reversibly intercalated in $SrLi_2Ti_6O_{14}$ and $BaLi_2Ti_6O_{14}$, while more than 4 lithium per formula unit (160 mAh/g) could be (de)inserted with distinct three-step profile in $PbLi_2Ti_6O_{14}$. Combustion made $MLi_2Ti_6O_{14}$ form a robust family of anode materials for Li-ion batteries.

Keywords Titanate · Anode · Battery

1 Introduction

Rechargeable lithium-ion batteries have single-handedly met energy needs of portable devices. Used as anode in (almost) all of the lithium-ion batteries because of its low-cost, high-capacity (372 mAh/g), and low-voltage (0.2 V vs lithium) layered graphite has critical issues. Owing to its low-voltage operation, they reduce electrolyte by stealing lithium from the cathode to form solid electrolyte interphase (SEI) and pose fire risk from dendrite short-circuit at times of fast charging or erroneous over-lithiation toward the end of discharge. Addressing these problems, titanate anodes led by zero-strain spinel $Li_4Ti_5O_{12}$ (LTO) [1] was introduced. Utilizing a Ti^{4+}/Ti^{3+} redox at ~1.5 V (vs Li), LTO does not form any SEI and is capable of high rate and safe operation. Following this, LTO was commercialized by Altair Nano, while Hydro-Quebec showed LTO—$LiFePO_4$ (LFP) 18,650 cells cycled at high rates (+15C/−5C) retaining more than 90% capacity for over 30,000 cycles [2]. Among numerous titanates investigated along these lines, orthorhombic *Cmca* $MLi_2Ti_6O_{14}$ (M = Sr, Ba, Pb) deliver voltage (1.3–1.45 V) and capacity (110–160 mA h/g) that

A. Chaupatnaik (✉) · P. Barpanda
Faraday Materials Laboratory, Materials Research Centre, Indian Institute of Science, Bengaluru 560012, India
e-mail: anshuman@iisc.ac.in

Y. Sharma et al. (eds.), *Recent Research Trends in Energy Storage Devices*,
https://doi.org/10.1007/978-981-15-6394-2_2

are comparable to LTO [3–5]. Further, as compared to LTO, they desirably have a lower area-specific resistance, working voltage, and higher packing density.

Since initial detection in form of single crystals grown using flux method, $MLi_2Ti_6O_{14}$ titanates have been prepared by sol-gel, template-assisted, electrospinning, and rheological phase reaction routes. Attempts largely by Jie Shu group in conventional solid-state reaction have been enriched with Ag coating, foreign element (Ag, Al, Ba, La, Mg, Na, N, Sr) doping/substitution, extending to inter-family solid solution. Most of the previous synthetic methods have used excess Li, stable TiO_2 precursor which mandates long ball-milling/annealing times. We report a two-step solution combustion method with fast annealing at low temperature [6–8]. Lithium-ion migration pathways have been shown in the structure and the battery activity (cyclic voltammetry and cycling stability) in lithium half-cell configuration are discussed.

2 Materials and Methods

Figure 1 shows a schematic for quick synthesis of $MLi_2Ti_6O_{14}$ by a two-step solution combustion method. While metal ions were stoichiometrically sourced from their respective metal nitrates (Sigma, >99%) dissolved in DI water, titanium (Ti^{4+}) ions were obtained from a metastable, thus, freshly prepared titanyl nitrate ($TiO(NO_3)_2$). This nitrate was prepared by nitric acid hydrolysis of titanium isopropoxide ($C_{12}H_{28}O_4Ti$, Sigma, 97%) in isopropyl alcohol (Merck, analytical

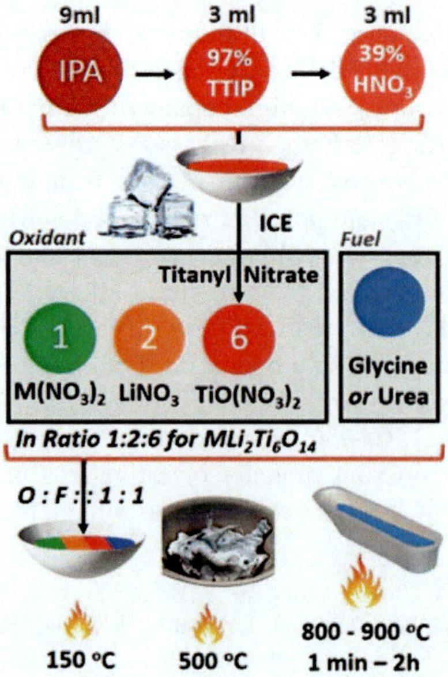

Fig. 1 Schematic diagram showing rapid synthesis of $MLi_2Ti_6O_{14}$ by solution combustion route. TTIP and IPA indicate titanium isopropoxide and isopropyl alcohol, respectively

grade 99.7%) at ice-cold conditions. These metal nitrates acting as oxidizers taken along with an equimolar amount of fuel (Glycine, Merck, 99.7%) in a ceramic bowl was burnt at 500 °C giving rise to white anthill-like formation. Following this, the ash was ground and rapidly annealed at 800 °C (2 h) for $BaLi_2Ti_6O_{14}$ and at 900 °C for 2 h and 1 min, respectively, in case of $SrLi_2Ti_6O_{14}$ and $PbLi_2Ti_6O_{14}$.

The phase purity of combustion prepared $MLi_2Ti_6O_{14}$ samples was assessed using powder X-ray diffraction patterns (2θ range = 10–50°, step size = 0.026° s^{-1}) acquired with a PANalytical X'Pert Pro X-ray diffractometer having a Cu-Kα source ($\lambda_1 = 1.54$ Å). Further, Rietveld refinement was done using the GSAS program with an EXPGUI graphical user interface [9]. Structures were illustrated using VESTA software. Post-gold coating (Quorum Q150R ES), particle, and element distribution in the titanate powders were analyzed using a Carl Zeiss ULTRA55 FESEM equipped with EDS system (Oxford Instruments). The structure and individual particle morphology were examined with an FEI Tecnai F30 STwin transmission electron microscope (TEM) operating at 200 kV. Pristine powders were dispersed in ethanol and drop cast on the carbon-copper grid (EMS) to prepare the TEM samples. The surface area was determined by Brunauer–Emmett–Teller (BET) method at 77 K by surface adsorption of N_2 (Belsorp Max).

The lithium-ion occupancy and their migration barriers in $MLi_2Ti_6O_{14}$ were studied using bond valence site energy (BVSE) method that has been extensively elaborated in previous works by Adam et al. [10]. The structural data fed form of refined atomic coordinates were obtained from the experiment. E_{BVSE}(Li) finds vacant interstitial regions of lowest BVSE and links them by intersite "bottleneck barrier" energies into pathways for migration of mobile lithium-ions. *soft*BV database [11] was used to obtain all required parameters.

The electrochemical battery performance the three titanates was checked in lithium half-cell format. Slurries with vacuum dried $MLi_2Ti_6O_{14}$ active material, polyvinylidene fluoride (PVDF) binder, and Super P carbon black were prepared in N-methyl-2-pyrrolidone (Sigma, >99) solvent in two formulations: 80:10:10 w/w for M = Sr, Ba and 70:20:10 w/w for M = Pb, respectively. To prepare electrodes, this slurry was uniformly drop cast on the copper current collector (Φ = 17 mm) and vacuum dried to eliminate NMP solvent. CR2032 coin cells were assembled inside an argon-filled glove box (MBraun LabStar GmbH). Coated disks and Li metal foils (Sigma Aldrich) were used as the working (cathode) and counter (anode) electrode, respectively, with a sheet of Whatman GF/C glass fiber used as the separator. 100 μl of 1 M $LiPF_6$ dissolved in 1:1:3 v/v % of ethylene carbonate/propylene carbonate/dimethyl carbonate (EC/PC/DMC) was dispensed as electrolyte (chameleon reagent). These coin cells were subjected to cyclic voltammetry at pt5 mV/s and pt1 mV/s in the voltage range of 0.1–3 V and 0.5–2 V. Cycle life test was done on separate cells at constant slow current (C/20, C/10, C/5) in the potential window of 0.5–2 V and 0.5–2.5 V (at 25 °C) with multi-channel BCS-810 (Bio-Logic SAS, France) battery cycler. The capacity (C) and current (C-rate) values were based on the theoretical capacity of $MLi_2Ti_6O_{14}$. These capacity values were 262 mAh/g, 243 mAh/g, and 220 mAh/g for M = Sr, Ba, and Pb, respectively.

3 Results and Discussion

3.1 *Synthesis*

The three $MLi_2Ti_6O_{14}$ titanates were prepared successfully by combustion synthesis. Urea was used as fuel in $SrLi_2Ti_6O_{14}$. In the case of M = Sr and Pb, attempts with different fuels converged on glycine. Glycine was used as it gave clear combustion flame and voluminous flaky product hinting high porosity and superior crystallinity. The $PbLi_2Ti_6O_{14}$ product formed using glycine fuel and rapid 1-min annealing was used to emphasize the short annealing time. Phase purity of the three-combustion synthesized $MLi_2Ti_6O_{14}$ titanates was confirmed from Rietveld refinement ($\chi 2 < 6.1$, R factor < 17.4%) (Figs. 2a–c) [12]. The lattice parameters of these orthorhombic (symmetry: *Cmca*) structures were similar to earlier reports. Calculated theoretical density (4.57 g/cm^3, 4.07 g/cm^3 for M = Pb and Ba) in the $MLi_2Ti_6O_{14}$ family (Sr = 3.85 g/cm^3) is higher than $Li_4Ti_5O_{12}$ (1 g/cm^{-3}) promising a high volumetric energy density anode. Global instability index (GII) values were calculated to be 0.12, 0.147, and 0.085 for M = Sr, Ba, and Pb that were found close to single crystal literature data of 0.10, 0.103, and 0.099. This validates the stability of the refined structure in terms of a stable chemical environment with reliable local structure and free volume

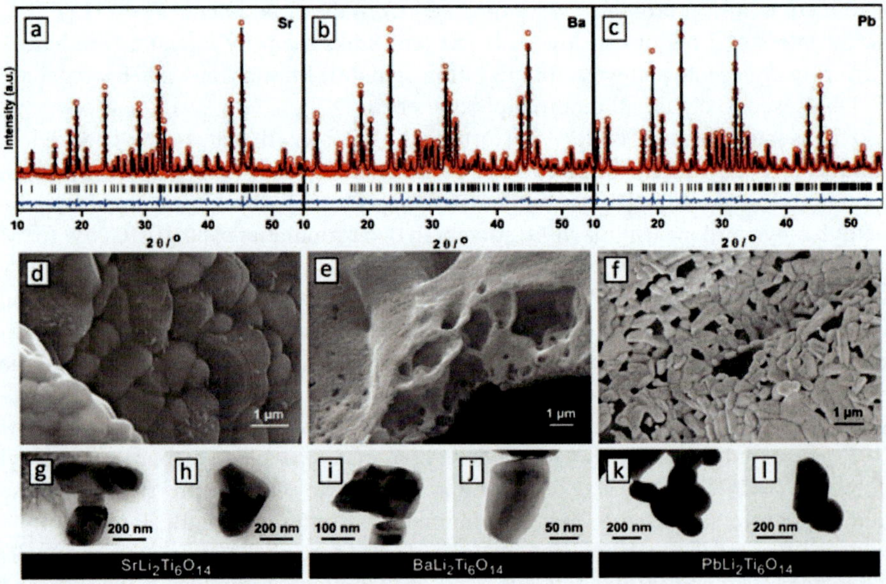

Fig. 2 Synthesis and particle distribution and morphology. **a–c** Rietveld refinement of XRD (Cu-Kα) patterns, **d** and **e** SEM micrograph, **g–l** TEM images of combustion prepared $SrLi_2Ti_6O_{14}$, $BaLi_2Ti_6O_{14}$, and $PbLi_2Ti_6O_{14}$, respectively. The experimental data points (red hollow dots), calculated pattern (black line), their difference (blue line) and Bragg reflections of the orthorhombic *Cmca* phases (black bars) are shown

needed later for BVSE analysis. A low bond valence sum (BVS) value for Sr (1.75) and Pb (1.79) indicates no residual strain at these sites unlike compressive strain at Ba site (BVS value 2.11) in case of $BaLi_2Ti_6O_{14}$.

Gas evolution and short annealing time ensured the formation of small primary nanometer-sized particles (100–300 nm, TEM) in the range of particle sizes only accessible by prolonged ball milling. TEM images (Fig. 2g–l) confirms nanometer-sized particles, which are smallest in case of $PbLi_2Ti_6O_{14}$. These primary particles coalesce to give secondary porous structures (SEM) resembling pomegranate seeds, cornflakes, and aerofoils for M = Sr, Ba, and Pb (Fig. 2d–f). The above morphology is expected to facilitate good electrolyte wetting while maximizing capacity utilization. The Brunner–Emmett–Teller (BET) specific surface area of the titanate powders was found to be in the range of 2–9 m^2/g.

3.2 *Structure*

As ionic movement is the slowest step and decides the highest operational C-rates, lithium diffusivity in $MLi_2Ti_6O_{14}$ structure has been studied. Figure 3 shows the

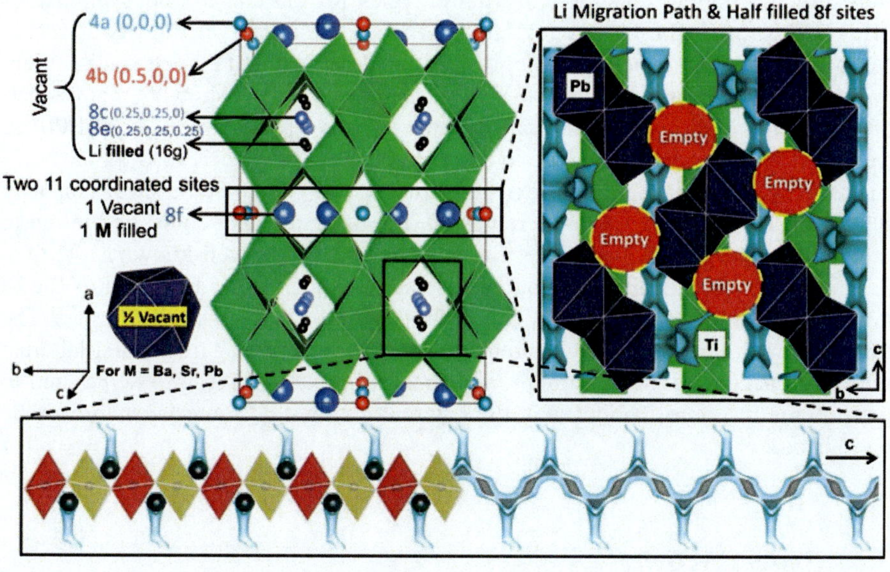

Fig. 3 Structure. VESTA illustration of $MLi_2Ti_6O_{14}$ structure. Vacant sites shown include empty 8c/8e sites (violet balls) in partly filled lithium (black ball) tunnel in a zig-zag manner, 4a/4b sites (red, light blue balls), and half of the 8f sites in the interlayer region. TiO_6 octahedra are represented in green. The right inset shows a *bc* section with lithium migration pathways (cyan tubes) along the *c*-axis, Pb filled 8f sites (blue polygon) and vacant 8f sites (red filled circles). Bottom inset shows the 1D migration path along the tunnels

structure of $MLi_2Ti_6O_{14}$ interlaced with lithium migration paths in the right/bottom inset. Structurally, a unit cell of $MLi_2Ti_6O_{14}$ is composed of Ti_2O_7 double layers in the *bc* plane "sandwiching" 11 (oxygen) coordinated M atoms placed in 8f positions in the interlayer mirror plane. These big 8f sites are fully occupied in case of 2Na but half-filled in case of M = Sr, Ba, and Pb, which can host 2 extra lithiums. Figure 3 (right inset) shows *bc* plan of the structure sliced at the mirror plane along with BVSE predicted Li^+ migration paths. The double layers consist of parallel *c*-axis directed rails of rutile TiO_2 held by two-set of TiO_2 rungs placed alternately. Guarded by these rungs and parallel to the rutile TiO_2 exist tunnels partly filled with Li. Inside the tunnel, structural lithium occupies tetrahedral 16 g sites in a zig-zag fashion (black balls in Fig. 3 bottom inset) while a chain of vacant octahedral 8c/8e sites lies in the core. 1D lithium migration pathways along *c*-direction connecting these empty core sites and structural lithium can be seen running parallel to the rutile TiO_2 chains. Additionally, in the interlayer mirror plane, vacant tunnels stitched by 4a/4b sites (Fig. 3) and empty half-filled 8f sites are expected to aid in lithium (de)insertion. BVSE give activation energies less than 1 eV for lithium migration along 1D, 2D, and 3D paths in case of $SrLi_2Ti_6O_{14}$ which exceeds 1 eV for 2D and 3D hops in case of M = Ba and Pb. Thus, $SrLi_2Ti_6O_{14}$ can be considered as a 3D ion conductor while $BaLi_2Ti_6O_{14}$ and $PbLi_2Ti_6O_{14}$ are limited to 1D ion conductors. The simple reason is an increasing volume trend of Sr–Pb–Ba which makes access to interstitial sites difficult, thereby raising activation energies for 2D and 3D hops in case of $MLi_2Ti_6O_{14}$ (M = Ba, Pb) having larger M ions. The interlayer threads of vacant 4a/4b sites have been phenomenologically considered without evidence for lithium occupancy in previous works. But BVSE rules them out by not showing them as part of any of the predicted 1D, 2D, or 3D paths in $MLi_2Ti_6O_{14}$ (M = Ba and Pb) as well as in a more appropriate case in relaxed structures of $MLi_{2+x}Ti_6O_{14}$ ($0 < x<6$), where additional x lithiums shift existing structural 16 g Li to 8c/8e sites. This may stem from BVSE ignoring Li–Li repulsion, but more likely it could be hinting at lithium insertion mechanism interestingly opposite to that of well-known $Li_4Ti_5O_{12}$. Summing up, while additional lithiums displace existing tetrahedral 8a Li to 16c Li in $Li_4Ti_5O_{12}$, lithiation in $MLi_2Ti_6O_{14}$ shifts 16 g lithiums to empty 8c/8e sites. The exact lithium insertion mechanism remains an open and exciting question. Elaborate efforts are needed to verify lithium movement in the tunnel (16 g to 8c/8e) and to find the reason, if so, behind why the external empty 4a/4b interlayer threads lie unlithiated.

3.3 Battery Activity

Reversible lithium storage in combustion prepared $MLi_2Ti_6O_{14}$ was analyzed in CR2032 lithium half cells using cyclic voltammetry and constant current cycling tests. During the first lithiation step, two independent things occur. First, lithium-ions insert into vacant sites as electrons reduce the structural Ti^{4+} to Ti^{3+}, second, as voltage falls below 1 V, electrons preferentially reduce the electrolyte on the surface

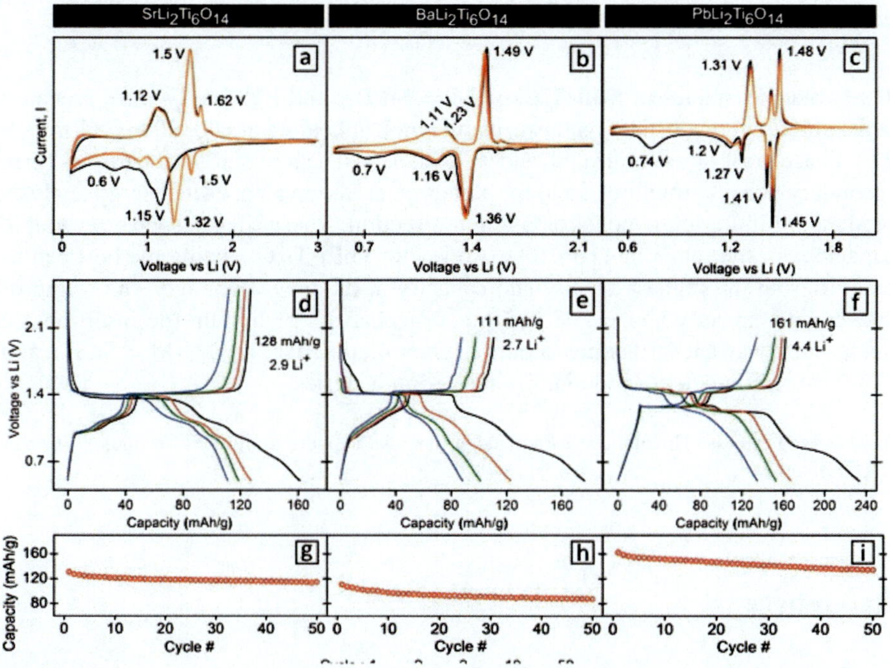

Fig. 4 Battery activity. **a–c** Cyclic voltammogram, **d–f** galvanostatic voltage profile, and **g–i** cycle life plots of $SrLi_2Ti_6O_{14}$, $BaLi_2Ti_6O_{14}$, and $PbLi_2Ti_6O_{14}$, respectively

of $MLi_2Ti_6O_{14}$ which together with incoming lithium forms stable solid electrolyte interphase (SEI). While the formation of SEI is evident from the first discharge reduction peaks of 0.6 V for M = Sr (Fig. 4a) and ~0.7 V for M = Ba and Pb (Fig. 4b, c), a change in the first discharge peaks compared to rest indicates local structure modification. M = Sr and Pb show three clear redox peaks while M = Ba shows two redox peaks with their respective values as shown in Fig. 4a–c. Cycle life tests were done at slow rate of C/20 for M = Sr and Ba and at fast rate of C/5 following a C/10 cycle in case of M = Pb. In the first charge, 128 mAh/g (2.9 Li) and 111 mAh/g (2.7 Li) could be extracted from $SrLi_2Ti_6O_{14}$ and $BaLi_2Ti_6O_{14}$. The capacity retention after 50 cycles was 88% and 78%, respectively. For $PbLi_2Ti_6O_{14}$, as much as 161 mAh/g (4.4 Li) could be extracted in the first cycle, 83% of which was retained after 50 cycles. The average voltage for lithium exchange was observed to be 1.38 V, 1.42 V, and 1.35 V, respectively, for M = Sr, Ba, and Pb.

4 Conclusion

Three titanate anodes in $MLi_2Ti_6O_{14}$ (M = Sr, Ba, and Pb) family were prepared using solution combustion synthesis using quick annealing at 800–900 °C (1 min—2 h). Phase purity was confirmed, and well-dispersed nanometer-sized particles form secondary porous structures ideal for battery use. Bond valence site energy analysis predicted lithium ion conduction along *c*-direction. These titanates gave reversible capacities in the range of 110–160 mAh/g with $PbLi_2Ti_6O_{14}$ being the best candidate offering the highest density and capacity at the lowest voltage. The origin of irreversible capacity above 1 V and exact mechanism of lithium (de)insertion are unclear and warrant further investigation. Overall, the $MLi_2Ti_6O_{14}$ (M = Sr, Ba, and Pb) forms a competent anode family for Li-ion batteries.

Acknowledgements The authors acknowledge financial support from Shell Technology Center (STC) Bangalore.

References

1. Ohzuku, T., Ueda, A., Yamamoto, N.: Zero-strain insertion material of $Li[Li_{1/3}Ti_{5/3}]O_4$ for rechargeable lithium cells. J. Electrochem. Soc. **142**(5), 1431–1435 (1995)
2. Zaghib, K., Dontigny, M., Guerfi, A., Charest, P., Rodrigues, I., Mauger, A., Julien, C.M.: Safe and fast-charging Li-ion battery with long shelf life for power applications. J. Power Sources **196**(8), 3949–3954 (2011)
3. Koseva, I., Chaminade, J.P., Gravereau, P., Pechev, S., Peshev, P., Etourneau, J.: A new family of isostructural titanates, $MLi_2Ti_6O_{14}$ (M = Sr, Ba, Pb). J. Alloys Compd. **389**(1–2), 47–54 (2005)
4. Belharouak, I., Amine, K.: $Li_2MTi_6O_{14}$ (M = Sr, Ba): new anodes for lithium-ion batteries. Electrochem. Commun. **5**(6), 435–438 (2003)
5. Li, P., Qian, S., Yu, H., Yan, L., Lin, X., Yang, K., Long, N., Shui, M., Shu, J.: $PbLi_2Ti_6O_{14}$: A novel high-rate long-life anode material for rechargeable lithium-ion batteries. J. Power Sources **330**, 45–54 (2016)
6. Dayamani, A., Shinde, G., Chaupatnaik, A., Rao, R.P., Adams, S., Barpanda, P.: Electrochemical and diffusional insights of combustion synthesized $SrLi_2Ti_6O_{14}$ negative insertion material for Li-ion Batteries. J. Power Sources **385**, 122–129 (2018)
7. Chaupatnaik, A., Barpanda, P.: Diffusional and electrochemical investigation of combustion synthesized $BaLi_2Ti_6O_{14}$ titanate anode for rechargeable batteries. J. Mater. Res. **34**(1), 158–168 (2018)
8. Chaupatnaik, A., Barpanda, P.: Swift combustion synthesis of $PbLi_2Ti_6O_{14}$ anode for lithium-ion batteries: diffusion and electrochemical investigation. J. Electrochem. Soc. **166**(3), A5122–A5130 (2019)
9. Larson, A.C., Von Dreele, R.B.: GSAS General Structure Analysis System. Los Alamos Natl. Lab. Rep. LAUR, 86–748 (2004)
10. Adams, S., Rao, R.P.: Transport pathways for mobile ions in disordered solids from the analysis of energy-scaled bond-valence mismatch landscapes. Phys. Chem. Chem. Phys. **11**, 3210–3216 (2019)

11. Chen, H., Adams, S.: Bond softness sensitive bond-valence parameters for crystal structure plausibility tests. IUCr J. **4**, 614–625 (2018)
12. Rietveld, H.M.: A profile refinement method for nuclear and magnetic structures. J. Appl. Crystallogr. **2**(2), 65–71 (1969)

Role of Annealing Temperature on Structural Modification of MoO_3 for Enhanced Electrochemical Properties

Arvind Singh, Bilal Ahmed, Saurav K. Ojha, and Animesh K. Ojha

Abstract Hexagonal phase of molybdenum trioxide (h-MoO_3) with rod-shaped structure was synthesized using solvothermal method at room temperature. The synthesized sample was further annealed at different temperatures. The annealed samples were characterized by various experimental techniques. Above 573 K, the hexagonal phase of MoO_3 is transformed to orthorhombic (α-MoO_3) along with the significant structural modifications (rod to sheet). The structural transformation induces defects, mainly oxygen vacancies, in the α-MoO_3. The α-MoO_3 may be formed due to the temperature-induced association and dissociation processes with the time. Cyclic voltammetry (CV) measurements of all the samples were done. The sample annealed at 673 K shows highest specific capacitance with the low value of charge transfer resistance along with good cycling stability. The better CV results of the sample annealed at 673 K were explained in terms of oxygen vacancies induced highly porous structure of α-MoO_3 compared to the other samples. The presence of abundant active sites at the surface and defects present inside the α-MoO_3 have made it as a better electrode material for its application in the field of energy storage devices.

Keywords Supercapacitor · MoO_3 · Annealing temperature

1 Introduction

For the past few decades, considerable attention has been made by the researchers towards the transition metal oxides, e.g., molybdenum trioxide (MoO_3), nickel oxide (NiO), vanadium oxide (V_2O_5), and RuO_2 as they provide a good platform for their applications in the field of energy storage [1–4]. Among these, MoO_3 is a n-type semiconductor and one of the adept materials due to its wide band gap, low-cost, large-scale production, smaller particle size, and large surface area and a suitable candidate to be used for energy storage. Till date, mainly three phases of MoO_3

A. Singh (✉) · B. Ahmed · S. K. Ojha · A. K. Ojha
Department of Physics, Motilal Nehru National Institute of Technology Allahabad, Prayagraj 211004, India
e-mail: singharvind8793@gmail.com

Y. Sharma et al. (eds.), *Recent Research Trends in Energy Storage Devices*,
https://doi.org/10.1007/978-981-15-6394-2_3

(hexagonal, monoclinic, and orthorhombic) were reported under different synthesis conditions [5, 6]. The simple topological arrangement of MoO_3 comprises of MoO_6 octahedron and connected distinctively for each polymorphic phase with the corner and edge sharing oxygen atoms. The phase transition from h-MoO_3 to α-MoO_3 probably exists with varying the synthesis conditions.

Among all the polymorphs, the stable α-MoO_3 is mainly studied due to its layered orthorhombic structure along (101) which causes to introduce oxygen vacancies [7]. The higher ion intercalation capability of MoO_3, which has MoO_6 octahedra with Vander wall gaps and a wide range of redox reactions, offers a new opportunity for supercapacitors [6]. However, MoO_3 has some shortcomings such as, normal reaction kinetics, low conductivity, and poor cycling behavior preclude the widespread use of energy storage applications. Recently, by the introduction of oxygen vacancies in MoO_3, the enhanced electrical conductivity resulting in improved electrochemical properties [7].

Herein, we report a simple route for the synthesis of MoO_3 by a cost-effective chemical method. The prepared sample was annealed at different temperatures and characterized using various techniques. The temperature-induced structural modifications may modify their properties and density of oxygen vacancy. We studied the role of temperature-induced structural modifications in MoO_3 on its supercapacitive performance.

2 Experimental

All the chemicals were used without further purification. The MoO_3 was synthesized by solvothermal method using ammonium molybdatetetrahydrate (AHM) $\{(NH_4)_6Mo_7O_{24}.4H_2O\}$, deionized water (DI), and nitric acid as precursors. Typically, 0.5 g AHM was added in 15 ml of DI water and stirred. The solution was heated at 343 K and added the 3 ml nitric acid into it while stirring. Finally, the precipitate was taken out and washed with ethanol followed by DI water to remove impurity and unreacted molecules. Further, the resultant sample was dried in the oven at 333 K for 12 h. The synthesized sample (M1) was annealed at various temperatures (473 K, 573 K, 673 K, 773 K, and 873 K) for 3 h. The samples annealed at 473 K, 573 K, 673 K, 773 K, and 873 K were denoted as M2, M3, M4, M5, and M6, respectively.

To explore electrochemical properties of the prepared samples, cyclic voltammetry (CV), galvanostatic charge-discharge (GCD) and electrochemical impedance spectroscopy (EIS) were carried out using a standard three-electrode system with the Metrohm Autolab PGSTAT302N, having Ag/AgCl (saturated KCl) as a reference electrode, platinum wire as counter electrode and a working electrode. The working electrodes were made-up by casting a mixture of active electrode material, carbon black, and polvinyidene difluoride (PVDF) binder, suspended in N-methyl-2-pyrrolidinone (NMP) in a mass ratio of 80:10:10 on Ni foam (area $1 \times 1\ cm^2$) and dried in air at 343 K. The electrochemical testing was done by taking 1 M H_2SO_4 as electrolyte at ambient temperature and pressure.

3 Results and Discussion

The X-ray diffraction (XRD) patterns of all the synthesized samples were shown in Fig. 1. From Fig. 1, it was confirmed that all the diffraction peaks of M1, M2, and M3 were well matched with the JCPDS card no. 21-0569 and having a single phase of h-MoO_3 [6]. There were no extra peaks appear in the XRD pattern that clearly indicates no impurity is present in the sample. However, as the temperature raised from 573 K, the phase of the samples was also changed. The main reflection peaks of M4, M5, and M6 can be indexed related to α-MoO_3, which is in good agreement with JCPDS card no. 35-0609 [8]. The smaller crystallite size of M4 indicating the larger surface area of the sample compared to other samples.

Figure 2 illustrated the scanning electron microscopy (SEM) images and energy-dispersive X-ray spectroscopy (EDS) spectra of M1 and M4. At room temperature, the synthesized h-MoO_3 is in the form of microrods (Fig. 2a) with average size 6–7 μm of length. The EDS spectra (Fig. 2b) shows only the peaks of Mo and O atoms. Since no other peaks were observed, it clearly indicates that no other phases were

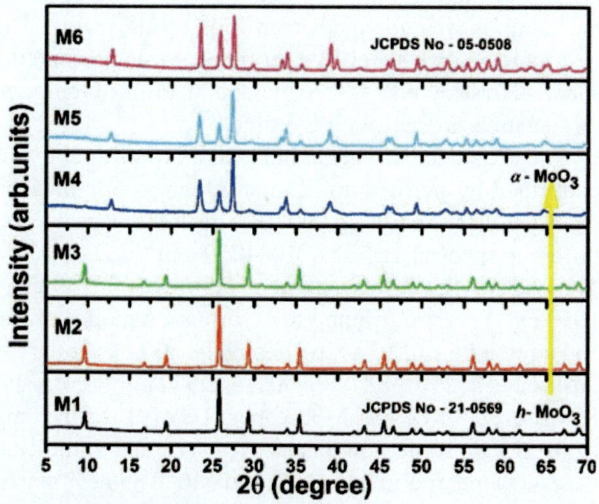

Fig. 1 XRD patterns of M1, M2, M3, M4, M5 and M6

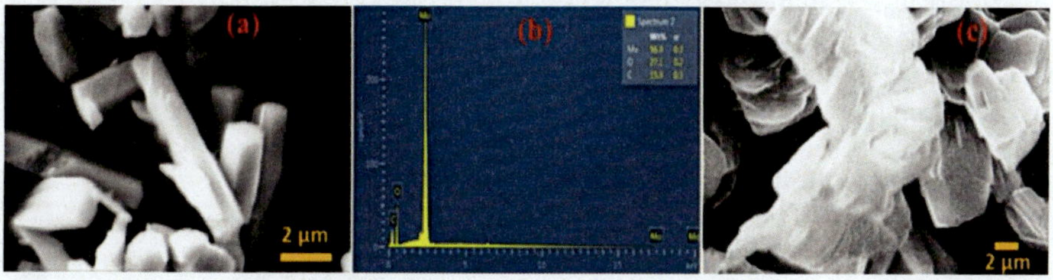

Fig. 2 SEM images with EDS spectra of M1 and M4

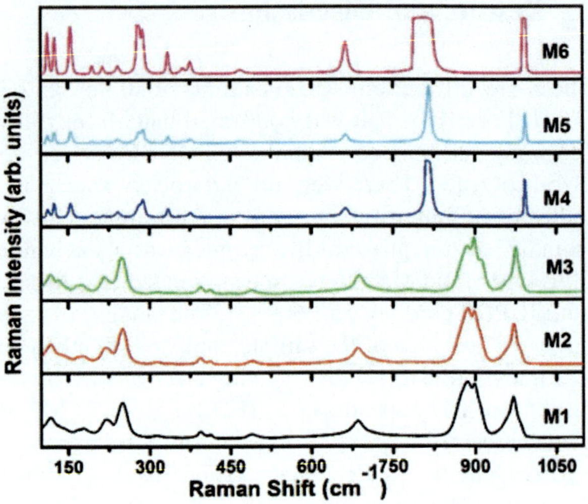

Fig. 3 Raman spectra of M1, M2, M3, M4, M5, and M6

present in the sample. As the temperature increases, the structural morphology of the samples was changed from rods to 2D sheets. The sample annealed at 673 K (α-MoO_3) shows sheet-like morphology as presented in Fig. 2c. The formation of sheets-like structures was explained in terms of temperature-induced association and dissociation process with the time.

The existence of hexagonal and orthorhombic phases of MoO_3 was further confirmed by performing Raman spectroscopy measurements and the spectra thus recorded were shown in Fig. 3. The Raman peaks were mainly observed in three different spectral regions, 100–250 cm^{-1}, 260–470 cm^{-1}, and 600–1050 cm^{-1}. The peaks lying in wavenumber regions 100–250 cm^{-1}, 260–470 cm^{-1}, and 600–1050 cm^{-1} were attributed to deformation modes of Mo–O–Mo, bending of Mo=O, and stretching of Mo–O, respectively [9–11]. As can be seen from Fig. 3, the peaks positioned at 250, 688, 885, 901, and 972 cm^{-1} correspond to the hexagonal phase of MoO_3 of M1, M2, and M3, respectively [9]. As the annealing temperature increases from 573 K, some new peaks appeared and some characteristic peaks of h-MoO_3 were disappeared in M4, M5, and M6. The new peaks at 243, 289, 338, 662, 813, and 989 cm^{-1} were belong to α-MoO_3. The strongest peak observed at ~813 cm^{-1} shift drastically with the change of phase from h-MoO_3 to α-MoO_3. This peak may be considered as signature peak to notify the phase transition from h-MoO_3 to α-MoO_3. Although, the extra peaks in α-MoO_3, may be due to the creation of oxygen vacancies on the surface. The narrow peak at 989 cm^{-1} was assigned to the terminal oxygen, stretching mode from unshared oxygen, authorized for the layered structure of α-MoO_3 which leads the creation of oxygen vacancies [11, 12].

4 Electrochemical Properties

To recognize the influence of phase and morphology on electrochemical properties, the samples were evaluated with the help of CV, GCD, and EIS measurements. CV measurements were carried out at different scan rates to study the specific capacitance of the synthesized samples in 1 M Na_2SO_4 electrolyte. Figure 4 displays the CV profile of the M4 sample at different scan rates. All CV curves exhibit a pair of redox peak at −0.09 V and −0.60 V vs. Ag/AgCl with the involvement of faradic redox reactions. Similar results were found in CV measurements of other samples of MoO_3. This indicates that the specific capacitance (SC) was governed by the adsorption and desorption of Na^+ ions on the surface of MoO_3 and this features evidently indication of the pseudocapacitive behavior of the material [13, 14] As the scan rate increases, the curves gradually deviated from the original shape which designates the degradation of electrochemical performance of the material [14]. Equation (1) is used to calculate SC of the sample using CV curves [15, 16].

$$\mathrm{SC} = \frac{1}{ms\Delta V}\int_{V1}^{V2} IdV \tag{1}$$

The inset of Fig. 4 shows the variation of SC with respect to the scan rate of all the samples. The results showed that sample M4 has a higher SC value of 415 F/g at 5 mV/s scan rate as compared to the other samples. The unique structural modifications (rods to sheets), phase, smaller crystallite size, resulting larger surface area of M4, which may be responsible for the enhancement in the value of SC.

To further estimate the SC value of all the samples, GCD tests were carried out. Figure 5 shows the GCD curve of M4 electrode in 1 M Na_2SO_4 electrolyte at different current densities. The nonlinear GCD curves showed the pseudocapacitance nature of the material [17]. Due to the resistive effect of the PVDF binder and low conductivity

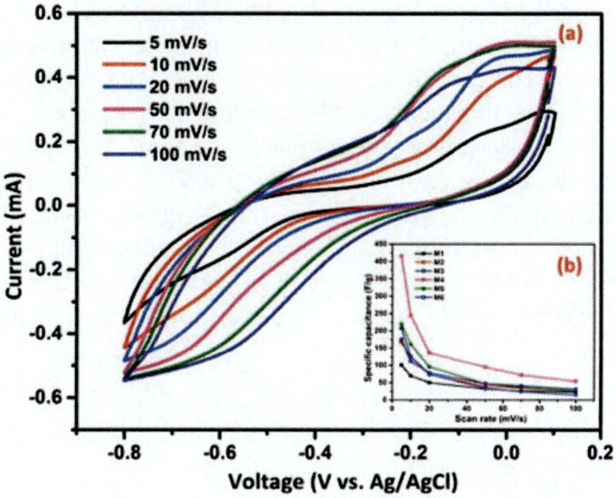

Fig. 4 **a** CV curves of M4 at different scan rates and inset **b** scan rate dependence of specific capacitance at different scan rate (in inset)

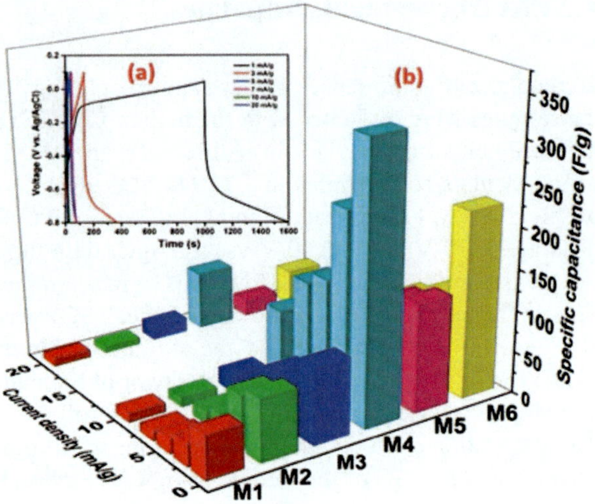

Fig. 5 **a** Galvanostatic charge-discharge curves of M4 and **b** specific capacitance of M1, M2, M3, M4, M5, and M6 with respect to current densities

of MoO_3, there is a large IR drop present in discharging profile [17]. The value of S_C was calculated using Eq. (2) by the GCD curves using the following relation correlated with discharging time (Δt), potential (ΔV), applied current density (I), and active mass of the electrode material (m). [18]

$$SC = \frac{I\Delta t}{m\Delta V} \tag{2}$$

The variation of SC with respect to current density of all the samples was shown in Fig. 5b. The sample M4 shows the highest SC among all the samples. The calculated values of SC of M4 were 332, 243, 142, 136, 82, and 73 F/g at current densities of 1, 3, 5, 7, 10, and 20 mA/g, respectively. The stability of any material is an important issue for device fabrication and therefore we check the electrochemical stability of M1 and M4 at the current density of 20 mA/s. The samples M1 and M4 retain ~96% and ~79% of SC over 3000 galvanostatic charge-discharge cycles (Fig. 6a). The decrease in capacitance with cycle in M4 may be due to the change of phases during redox reactions and low electron transfer kinetics [7, 15].

EIS measurement was done in the frequency range of 0.01 Hz to 100 kHz with an alternating current (AC) of 10 mV to identify the charge transfer behavior of the materials. Figure 6b shows the Nyquist plots of M1 and M4 samples. From the graphs (inset magnified image), it is clear that sample M4 has lower charge transfer resistance in comparison with M1 [19]. At low frequency, the material showed the capacitive nature that was further supported by CV and GCD measurements.

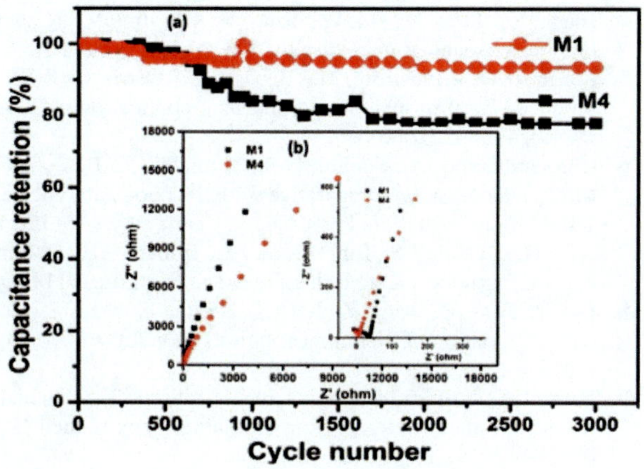

Fig. 6 **a** Stability profiles of M1 and M4 **b** Nyquist plots inset: magnified images of Nyquist plots

5 Conclusion

In this study, h-MoO_3 was synthesized by cost-effective, large-scale production by solvothermal method. The sample is annealed at different temperatures in air for 3 h to get the α-MoO_3 crystalline phase. The phase and morphology of the samples were analyzed by XRD and SEM measurements. The presence of oxygen vacancy is further confirmed by the Raman analysis. The sample annealed at 673 K demonstrated the highest SC of 415 F/g at the 5 mV/s scan rate compared to other samples. The structural modification and oxygen vacancies play an important role for enhancing the electrochemical properties of the material. Cycling stability of the material can be improved by preventing the phase transition during charge-discharge process. Thus, the present work demonstrates the electrochemical storage properties of h-MoO_3 and α-MoO_3.

Acknowledgements The authors are grateful to MNNIT Allahabad, Prayagraj and TEQIP-III for Financial supports. The instrumentation facilities available in the Center for Interdisciplinary Research (CIR) of the Institute are greatly acknowledged.

References

1. Pujari, R.B., Lokhande, V.C., Kumbhar, V.S., Chodankar, N.R., Lokhande, C.D.: Hexagonal microrods architectured MoO_3 thin film for supercapacitor application. J. Mater. Sci.: Mater. Electron. **27**(4), 3312–3317 (2016)
2. Ren, X., Guo, C., Xu, L., Li, T., Hou, L., Wei, Y.: Facile synthesis of hierarchical mesoporous honeycomb-like NiO for aqueous asymmetric supercapacitors. ACS Appl. Mater. Interfaces **7**(36), 19930–19940 (2015)
3. Subramanian, V., Hall, S.C., Smith, P.H., Rambabu, B.: Mesoporous anhydrous RuO_2 as a supercapacitor electrode material. Solid State Ionics **175**(1–4), 511–515 (2004)

4. Liang, K., Tang, X., Hu, W., Yang, Y.: Ultrafine V_2O_5 nanowires in 3D current collector for high-performance supercapacitor. ChemElectroChem **3**(5), 704–708 (2016)
5. Mariotti, D., Lindström, H., Bose, A.C., Ostrikov, K.K.: Monoclinic β-MoO_3 nanosheets produced by atmospheric microplasma: application to lithium-ion batteries. Nanotechnology **19**(49), 495302 (2008)
6. Chithambararaj, A., Sanjini, N.S., Velmathi, S., Bose, A.C.: Preparation of h-MoO_3 and α-MoO_3 nanocrystals: comparative study on photocatalytic degradation of methylene blue under visible light irradiation. Phys. Chem. Chem. Phys. **15**(35), 14761–14769 (2013)
7. Kim, H.S., Cook, J.B., Lin, H., Ko, J.S., Tolbert, S.H., Ozolins, V., Dunn, B.: Oxygen vacancies enhance pseudocapacitive charge storage properties of MoO_{3-x}. Nat. Mater. **16**(4), 454 (2017)
8. Liu, Y., Feng, P., Wang, Z., Jiao, X., Akhtar, F.: Novel fabrication and enhanced photocatalytic MB degradation of hierarchical porous monoliths of MoO_3 nanoplates. Sci. Rep. **7**(1), 1845 (2017)
9. Yano, T.A., Yoshida, K., Hayamizu, Y., Hayashi, T., Ohuchi, F., Hara, M. Probing edge-activated resonant Raman scattering from mechanically exfoliated 2D MoO_3 nanolayers. 2D Mater. **2**(3), 035004 (2015)
10. Hirata, T., Ishioka, K., Kitajima, M.: Raman spectra of MoO_3 implanted with protons. Appl. Phys. Lett. **68**(4), 458–460 (1996)
11. Singh, A., Kumar, S., Ahmed, B., Singh, R. K., Ojha, A.K.: Temperature induced modifications in shapes and crystal phases of MoO_3 for enhanced photocatalytic degradation of dye waste water pollutants under UV irradiation. J. Alloys Compd. **806**, 1368–1376 (2019)
12. Sharma, R.K., Reddy, G.B.: Synthesis and characterization of α-MoO_3 microspheres packed with nanoflakes. J. Phys. D Appl. Phys. **47**(6), 065305 (2014)
13. Kumar, V., Lee, P.S.: Redox active polyaniline-h-MoO_3 hollow nanorods for improved pseudocapacitive performance. J. Phys. Chem. C **119**(17), 9041–9049 (2015)
14. Chen, J., Han, S., Zhao, H., Bai, J., Wang, L., Sun, G., Xie, E.: Robust wire-based supercapacitors based on hierarchical α-MoO_3 nanosheet arrays with well-aligned laminated structure. Chem. Eng. J. **320**, 34–42 (2017)
15. Mahmood, Q., Yun, H.J., Kim, W.S., Park, H.S.: Highly uniform deposition of MoO_3 nanodots on multiwalled carbon nanotubes for improved performance of supercapacitors. J. Power Sources **235**, 187–192 (2013)
16. Nagarajan, N., Humadi, H., Zhitomirsky, I.: Cathodic electrodeposition of MnO_x films for electrochemical supercapacitors. Electrochim. Acta **51**(15), 3039–3045 (2006)
17. Prakash, N.G., Dhananjaya, M., Narayana, A.L., Shaik, D.P., Rosaiah, P., Hussain, O.M.: High performance one dimensional α-MoO_3 nanorods for supercapacitor applications. Ceram. Int. **44**(8), 9967–9975 (2018)
18. Hu, C.C., Tsou, T.W.: Ideal capacitive behavior of hydrous manganese oxide prepared by anodic deposition. Electrochem. Commun. **4**(2), 105–109 (2002)
19. Khandare, L., Late, D.J.: MoO_3-rGO nanocomposites for electrochemical energy storage. Appl. Surf. Sci. **418**, 2–8 (2017)

Reversible Sodium and Potassium-Ion Intercalation in $Na_{0.44}MnO_2$

Baskar Senthilkumar, Sai Pranav Vanam, and Prabeer Barpanda

Abstract Global demands drew attention towards energy production and storage technologies in the twenty-first century. Lithium-ion batteries cannot satisfy needs because of limited resources of Li. Sodium-ion batteries can be a promising alternate replacement with a redox potential E (Na^+/Na) = −2.71 V vs SHE. Potassium-ion batteries have recently gained interest in competing with sodium with a redox potential of E (K^+/K) = −2.93 V and has an advantage of smaller stokes radius. Herein, we tried to understand the K-insertion and de-insertion in sodium-based transition metal oxide ($Na_{0.44}MnO_2$). The tunnel-type $Na_{0.44}MnO_2$ was prepared by combustion synthesis technique which has an orthorhombic structure with a space group of *Pbam*. It has 3D tunnel-type structure with large and small tunnels as a path for diffusion of ions. For the first time, tunnel-type $Na_{0.44}MnO_2$ was investigated as K^+-ion intercalation host. This study shows reversible K^+-ion intercalation with an initial discharge of 74 mAh g^{-1} at C/20 current rate and good cycling stability up to 50 cycles.

Keywords Sodium manganese oxide · Cathodes · Potassium-ion battery

1 Introduction

As the demand for large-scale energy storage devices is growing tremendously, lithium-ion battery technology can not fulfil all the needs. The expanding commerical applications like electric vehicles and grid-level storage raise concern over the availability of ample lithium in the earth's crust. In the quest for alternative storage devices, the nearest elements like Na and Mg have come to focus from past few decades because of similar electrochemical behaviour. Na-ion batteries have been extensively investigated focusing on variety of transition metal oxides [1, 2] and polyanionic materials [3, 4]. Even though nearest neighbour elements are the best choices, K is indeed giving a competition to Na in the sections of abundance, good

B. Senthilkumar (✉) · S. P. Vanam · P. Barpanda
Faraday Materials Laboratory, Materials Research Centre, Indian Institute of Science, Bengaluru 560012, India
e-mail: baskars@iisc.ac.in

Y. Sharma et al. (eds.), *Recent Research Trends in Energy Storage Devices*,
https://doi.org/10.1007/978-981-15-6394-2_4

anode, and cathode materials. Even more, K-intercalation is more feasible in graphite (anode material) than in case of sodium due to low solvation radius [5]. In comparing the redox potentials (vs SHE) with Li/Li^+($= -3.04$ V) and Na/Na^+($= -2.71$ V), K/K^+($= -2.93$ V) stanin between giving an added advantage.

In growing attraction of K-ion batteries (KIBs), potassium (de)intercalations have been studied for various materials. The first reported cathode material was Prussian blue by Eftekhari [6] in 2004 showing a capacity of 77 mAh g^{-1} at C/10 current rate and capacity loss of only 12% even after 500 cycles. In recent years, various cathode materials were investigated which includes layered metal oxides [7, 8], polyanionic compounds [9], and also organic materials [10]. In cobalt oxide-based system, K_xCoO_2 was reported by Komaba et al. [11] (half-cell) and Ceder et al. [12] (full cell) delivering a reversible capacity of 60 and 53 mAh g^{-1}, respectively. Shortly, after that, Sada et al. [13] have reported reversible capacity of 82 mAh g^{-1} when used as a working electrode as Na_xCoO_2. Later, many Na-based oxide materials were studied like in $Na_{0.52}CrO_2$ by Myoungho et al. [14] with a capacity of 88 mAh g^{-1}.

More recently, Na–Mn–O-based ternary compound was studied for K (de)intercalation by Sada et al. [15] showing an excellent capacity of 140 mAh g^{-1}. The reports on Na-based metal oxide as intercalation host for K^+ ions motivated us to investigate the well-known tunnel-type $Na_{0.44}MnO_2$ cathode material. The present study focuses on using $Na_{0.44}MnO_2$ cathode material as K-intercalation host. Mn-based materials have been extensively studied as positive electrode materials for Li and Na-ion batteries due to its environmental friendly, abundance, non-toxicity, and thermal stability. $Na_{0.44}MnO_2$ has an advantage of wide tunnel structure feasible for sodium/potassium insertion/de-insertion. It is thermodynamically stable in between the sodium content in the system which is in the range of 0.22–0.66. Owing to high theoretical capacity of 121 mAh g^{-1}, $Na_{0.44}MnO_2$ is a promising positive electrode material for Na(K)-ion batteries. This work is an effort to investigate the reversible K^+ (de)intercalation in this material.

2 Materials and Methods

$Na_{0.44}MnO_2$ was synthesised by using simple solution combustion synthesis (SCS) method. At first, stoichiometric amounts of sodium nitrate ($NaNO_3$) and manganese nitrate ($Mn(NO_3)_2$) were dissolved in distilled water. Excess (5 wt%) of $NaNO_3$ were added to compensate Na loss at high temperature annealing. Urea was used as a fuel and its amount was maintained with oxidiser to fuel ratio (f) to be 1. The stirred mixture was maintained at 100 °C on a hot plate, which was later raised slowly to 200 °C to trigger the combustion reaction. The ash obtained after combustion was collected, mildly grinded and later calcined at 900 °C for 3 h with the heating time of 5 °C/min. Phase conformation of calcined powder was characterized by X-ray powder diffraction with PANalytical X'pert pro diffractometer (source: Cu Kα operating at 40 kV and 30 mA). Rietveld refinement was carried out using FullProf.

Sample morphology was verified using field emission scanning electron microscopy (FE-SEM), Ultra 55 FESEM Carl Zeiss EDS, operated at 5–20 kV.

The electrodes for electrochemical characterization were prepared by mixing active material ($Na_{0.44}MnO_2$), super P-carbon black, and polyvinylidene fluoride (PVDF) binder in 80:10:10 (w/w/w) ratios in N-methyl pyrrolidone (NMP) solvent. The resultant slurry was coated on the Al-foil and dried overnight at 80 °C in vacuum. Electrochemical test was performed by using a CR2032-type coin cell. The active mass loading in each working electrode was ~2 mg cm^{-2}. The assembling of coin cell was done in glove box with H_2O and O_2 level <0.5 ppm. Metallic Na(K) foil was used as counter electrode for Na-ion (K-ion) half-cells. Glass microfibre filter (Whatman, grade GF/C) was used as separator. The electrolytes used for Na-ion and K-ion half-cells were 1 M $NaPF_6$ and 0.8 M KPF_6 in ethylene carbonate (EC)/diethyl carbonate (DEC) (1:1 (v/v)), respectively. Galvanostatic charge/discharge test was conducted in the voltage range of 4.0–2.0 V for Na-ion half-cell and 4.2–2.0 V for K-ion half-cell.

3 Results and Discussion

XRD pattern of combustion synthesized $Na_{0.44}MnO_2$ is shown in Fig. 1a. The characteristic diffraction peaks can be indexed to orthorhombic structure with *Pbam* space group (JCPDS card no. 27-0750). Rietveld refinement result confirms crystallinity of the compound and the refined lattice parameters, $a = 9.086(8)$ Å; $b = 26.467(8)$ Å; $c = 2.825(3)$ Å, are in good agreement with the previous reports [16, 17]. As schematic illustration of crystal structure shown in Fig. 1b, the framework of $Na_{0.44}MnO_2$ is built with the MO_6 octahedra and MO_5 pyramids. These two kinds of polyhedra are corner shared whereas the MO_6 are edge shared giving rise to two

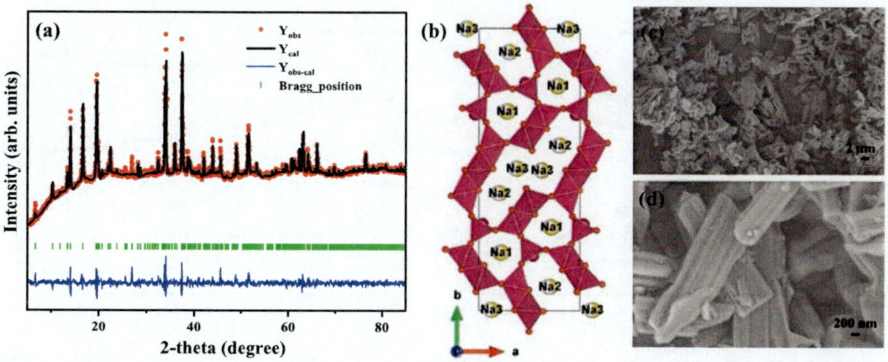

Fig. 1 **a** Rietveld refinement result of $Na_{0.44}MnO_2$ using FullProf; **b** 3D tunnel like structure shows Na1, Na2, Na3; **c**, **d** SEM images shown in lower and higher magnification

different tunnels. There are three Na atom sites in the structure named as Na1, Na2, and Na3 as shown in Fig. 1b. Na2 and Na3 are situated in large S-shaped tunnels that are fully occupied. On the other hand, Na1 is placed in smaller tunnels. These atoms are feasible for going in and out of the system which occurs mainly along the c-axis. Images observed by SEM as shown in Fig. 1c, d display a rod-like morphology. The width of rods is around 1 μm and the length is ~5 to 10 μm. Due to high temperature calcination, the rods are bundled.

Figure 2a, c plots represent the galvanostatic charge/discharge profiles of $Na_{0.44}MnO_2$ in Na-ion half-cell recorded at two different current rates of C/20 and 1C. The charge/discharge profile recorded at C/20 shows a capacity of over 120 mAh g^{-1} (Fig. 2a). The first discharge capacity is higher than first charge, indicating more amount of Na^+ ions are inserted into the structure. As reported by Sauvage et al., 0.22 Na per unit cell can be extracted in the initial charge that is clearly observed in the charge profile of Fig. 2a. During first discharge, 0.44 Na per unit cell can be inserted back into the structure resulting a total Na content of 0.66. The differential capacity (dQ/dV) of C/20 shows six voltage plateaus indicating multiphase transition during sodiation and de-sodiation process [16]. First principle calculations show Na1 site could also be inserted which gives contributions to the capacity [18]. The material

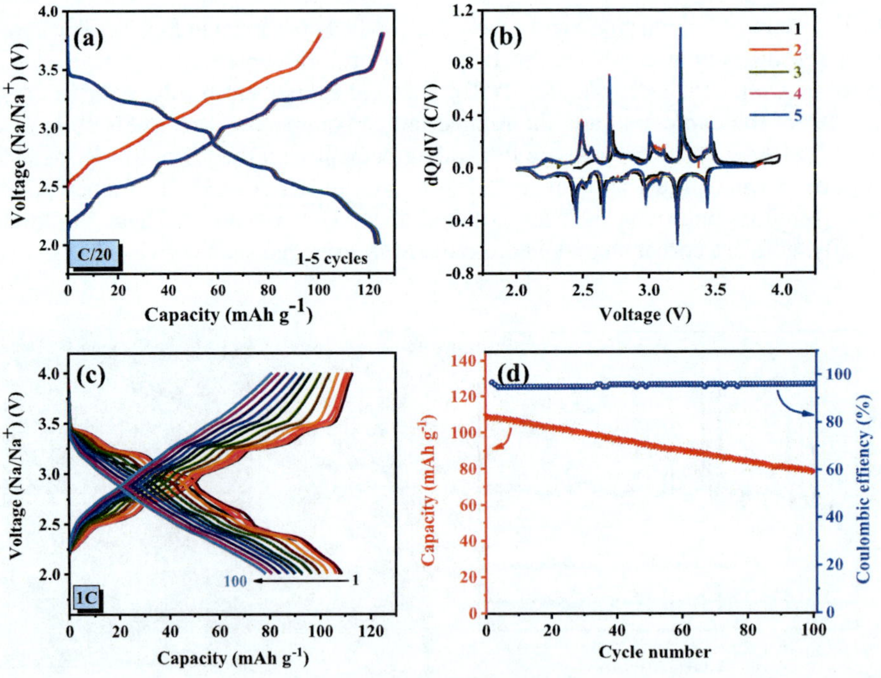

Fig. 2 Galvanostatic charge/discharge profile of $Na_{0.44}MnO_2$ in a Na-ion half-cell architecture **a** at C/20 current rate; **c** at 1C current rate; **b** differential capacity (dQ/dV) curves of cycles 1–5 at current rate C/20; **d** cycling performance at 1C

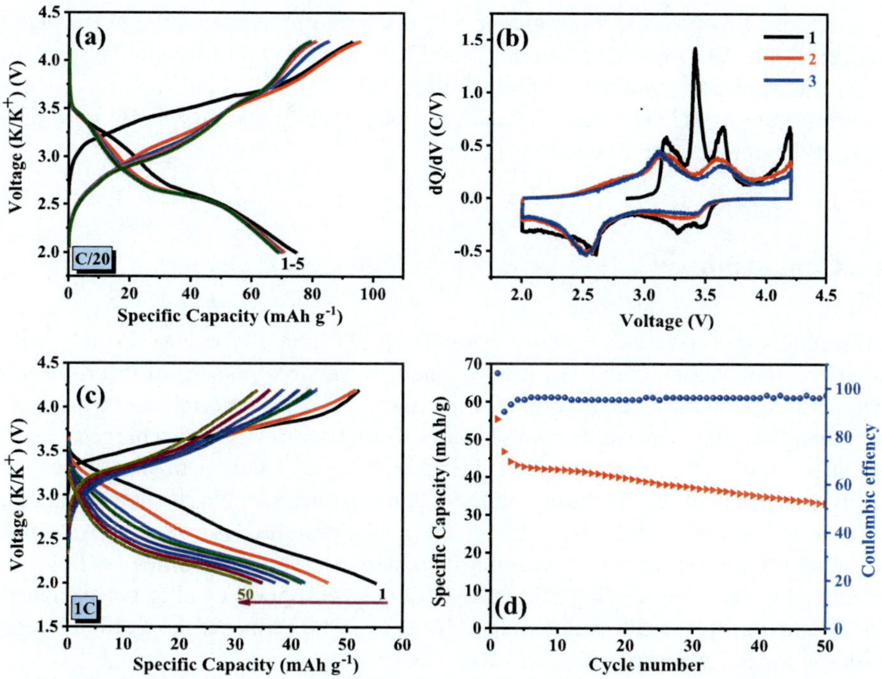

Fig. 3 Galvanostatic charge/discharge profile of $Na_{0.44}MnO_2$ in a K-ion half-cell architecture **a** at C/20 current rate; **c** at 1C current rate; **b** differential capacity (dQ/dV) curves of cycles 1–5 at current rate C/20; **d** cycling performance at 1C

is tested at high current rate (1C) up to 100 cycles, and gives a capacity retention of 70% which is 80 mAh g^{-1} even after 100 cycles.

The K-ion intercalation behaviour in $Na_{0.44}MnO_2$ was studied in the voltage range of 2.0–4.2 V. Figure 3a represents the galvanostatic charge/discharge profiles of $Na_{0.44}MnO_2$ in K-ion half-cell recorded at a current rate of C/20. First, the K-ion cell was charged up to 4.2 V and then discharged to 2 V at a constant current rate of C/20 (Fig. 3a). During the first charge, the Na ion starts de-intercalating from the tunnel-type $Na_{0.44}MnO_2$ structure and it provided an initial charge capacity of ~100 mAh g^{-1}. During the discharge, K-ions are inserted into the vacant Na sites. This can be observed from variation in the discharge profile as compared to sodiation in Fig. 2a. According to the maximum potassium insertion of 0.44 K^+ and complete exchange of Na by K, theoretical capacity is calculated to be 113 mAh g^{-1}. However, we observed an initial discharge of 74 mAh g^{-1} at C/20 (Fig. 3a) which is close to 0.29 K^+-ion insertion per f.u.

With capacity loss in the initial two cycles, the voltage profiles of subsequent cycles were completely overlapping (Fig. 3a) proving efficient reversibility. From differential capacity (dQ/dV) curves in Fig. 3b, it is clearly seen that during first discharge three peaks are seen, and this shows a structural change. In the later cycles, only two peaks in both charge and discharge indicating structure is stable and reversible. The detailed exsitu XRD and XPS studies will be reported elsewhere. Figure 3c shows the voltage profiles of $Na_{0.44}MnO_2$ in K-ion half-cell recorded at a current rate of 1C. The material showed good cycling stability at 1C with 63% capacity retention after 50 cycles (Fig. 3d).

4 Conclusion

In summary, we have successfully synthesised 3D tunnel-type $Na_{0.44}MnO_2$ using solution combustion synthesis (SCS) method. Reversible potassium insertion into $Na_{0.44}MnO_2$ host has been reported for the first time. This material can be used as a promising cathode material for both Na and K-ion batteries. The discharge capacity of Na-ion half-cell and K-ion half-cell is 123 mAh g^{-1} and 74 mAh g^{-1}, respectively. Electrochemical oxidation and reduction were used for Na-ion deintercalation followed by K-ion intercalation. The resulting host structure $Na_{0.22}K_xMnO_2$ showed reversible K-ion de-insertion involving two stepwise voltage profiles leading to a discharge capacity exceeding 70 mAh g^{-1} along with good cycling performances (63% retention up to 50 cycles). $Na_{0.44}MnO_2$ can be proposed as a suitable host material for K-ion insertion.

Acknowledgements The authors thank the financial support from the Science and Engineering Research Board (SERB, Government of India) in form of an Early Career Research Award (ECR/2015/000525). B.S. thanks SERB for a National Postdoctoral Fellowship (PDF/2015/00217). The authors thank Shell Technology Centre (STC) Bangalore for financial support.

References

1. Liu, Q., Hu, Z., Chen, M., Gu, Q., Dou, Y., Sun, Z., Chou, S., Dou, S.X.: Multiangular rod-shaped $Na_{0.44}MnO_2$ as cathode materials with high rate and long life for sodium-ion batteries. ACS Appl. Mater. Interfaces **9**(4), 3644–3652 (2017)
2. Chen, Z., Yuan, T., Pu, X., Yang, H., Ai, X., Xia, Y., Cao, Y.: Symmetric sodium-ion capacitor based on $Na_{0.44}MnO_2$ nanorods for low-cost and high-performance energy storage. ACS Appl. Mater. Interfaces **10**(14), 11689–11698 (2018)
3. Nose, M., Nakayama, H., Nobuhara, K., Yamaguchi, H., Nakanishi, S., Iba, H.: $Na_4Co_3(PO_4)_2P_2O_7$: A novel storage material for sodium-ion batteries. J. Power Sources **234**, 175–179 (2013)
4. Park, Y.U., Seo, D.H., Kim, H., Kim, J., Lee, S., Kim, B., Kang, K.: A family of high-performance cathode materials for Na-ion batteries, $Na_3(VO_{1-x}PO_4)_2F_{1+2x}$ $(0 \le x \le 1)$: combined first-principles and experimental study. Adv. Funct. Mater. **24**(29), 4603–4614 (2014)
5. Komaba, S., Hasegawa, T., Dahbi, M., Kubota, K.: Potassium intercalation into graphite to realize high-voltage/high-power potassium-ion batteries and potassium-ion capacitors. Electrochem. Commun. **60**, 172–175 (2015)
6. Eftekhari, A.: A. Potassium secondary cell based on prussian blue cathode. J. Power Sources **126**(1), 221–228 (2004)

7. Vaalma, C., Giffin, A.G., Buchholz, D., Passerini, S.: Non-aqueous K-ion battery based on layered $K_{0.3}MnO_2$ and hard carbon/carbon black. J. Electrochem. Soc. **163**, 1295–1299 (2016)
8. Wang, X., Xu, X., Niu, C., Meng, J., Huang, M., Liu, X., Liu, Z., Mai, L.: Earth abundant Fe/Mn-based layered oxide interconnected nanowires for advanced K-ion full batteries. Nano Lett. **17**, 544–550 (2017)
9. Chihara, K., Katogi, A., Kubota, K., Komaba, S.: $KVPO_4F$ and $KVOPO_4$ toward 4 volt-class potassium-ion batteries. Chem. Commun. **53**, 5208–5211 (2017)
10. Xing, Z., Jian, Z., Luo, W., Qi, Y., Bommier, C., Chong, E.S., Li, Z., Hu, L., Ji, X.: A perylene anhydride crystal as a reversible electrode for K-ion batteries. Energy Storage Mater. **2**, 63–68 (2016)
11. Hironaka, Y., Kubota, K., Komaba, S.: P2-and P3-K_xCoO_2 as an electrochemical potassium intercalation host. Chem. Commun. **53**, 3693–3696 (2017)
12. Kim, H., Kim, J.C., Bo, S.H., Shi, T., Kwon, D.H., Ceder, G.: K-ion batteries based on a P2-type $K_{0.6}CoO_2$ cathode. Adv. Energy Mater. **7**(17), 1700098 (2017)
13. Sada, K., Senthilkumar, B., Barpanda, P.: Electrochemical potassium-ion intercalation in Na_xCoO_2: a novel cathode material for potassium-ion batteries. Chem. Commun. **53**(61), 8588–8591 (2017)
14. Naveen, N., Woon, B.P., Su, C.H., Satendra, P.S., Young, H.J., Docheon, A., Kee, S.S., Myoungho, P.: Reversible K^+-insertion/deinsertion and concomitant Na^+-redistribution in P′3-$Na_{0.52}CrO_2$ for high-performance potassium-ion battery cathodes. Chem. Mater., **30**(6), 2049–2057 (2018)
15. Sada, K., Senthilkumar, B., Barpanda, P.: Potassium-ion intercalation mechanism in layered $Na_2Mn_3O_7$. ACS Appl. Energy Mater. **1**(10), 5410–5416 (2018)
16. Sauvage, F., Laffont, L., Tarascon, J.M., Baudrin, E.: Study of the insertion/deinsertion mechanism of sodium into $Na_{0.44}MnO_2$. Inorg. Chem. **46**(8), 3289–3294 (2007)
17. Wang, C.H., Yeh, Y.W., Wongittharom, N., Wang, Y.C., Tseng, C.J., Lee, S.W., Chang, W.S., Chang, J.K.: Rechargeable Na/$Na_{0.44}MnO_2$ cells with ionic liquid electrolytes containing various sodium solutes. J. Power Sources **274**, 1016–1023 (2015)
18. Heejin, K., Dong, J.K., Dong, H.S., Min, S.Y., Kisuk, K., Do, K.K., Yousung, J.: Ab Initio study of the sodium intercalation and intermediate phases in $Na_{0.44}MnO_2$ for sodium-ion battery. Chem. Mater. **24**(6), 1205–1211 (2012)

Cobalt–Phosphate-Based Insertion Material as a Multifunctional Cathode for Rechargeable Hybrid Sodium–Air Batteries

Chinnasamy Murugesan, Baskar Senthilkumar, Kriti Choudhary, and Prabeer Barpanda

Abstract Herein, we introduce the concept of using multifunctional air–cathode to advance the performance of hybrid sodium–air batteries (NABs) by utilizing intercalation/deintercalation followed by OER/ORR mechanism. As a case study, sodium tetra-cobalt triphosphate $NaCo_4(PO_4)_3$ was chosen and was synthesized by solution combustion synthesis technique. The synthesised $NaCo_4(PO_4)_3$ had unique 5-coordination cobalt, which offers a different environment to the incoming oxygen that facilitates the oxygen-binding on the active site. It showed porous morphology that allows proper wettability of the electrolyte. The $NaCo_4(PO_4)_3$ exhibits bi-functional OER/ORR catalytic activity as well as OH^- ion (de)intercalation due to its multiple redox behaviours. The possible ORR and OER mechanism is explored. The hybrid NABs with $NaCo_4(PO_4)_3$ as air–cathode delivered high power and energy density with improved round-trip efficiency.

Keywords Na–air battery · Cobalt–phosphate · Cathode

1 Introduction

Metal–air batteries are essential for sustainable growth of using renewable energy resources. Developing cost-effective sodium–air batteries (NABs) is considered an innovative approach to find high energy battery device for stationary grid-storage applications [1]. Precious metal-based catalysts (Pt, Ir and Ru) are known for superior performance. However, their widespread use has been limited by their availability and cost. To address this issue, many non-precious carbon-based materials have been proposed as a preferred bi-functional catalyst. The lack of long-term durability and

C. Murugesan (✉) · B. Senthilkumar · P. Barpanda
Faraday Materials Laboratory, Materials Research Centre, Indian Institute of Science, Bengaluru 560012, India
e-mail: chinnasamym@iisc.ac.in

K. Choudhary
Department of Chemistry, Indian Institute of Technology, Roorkee, Uttarakhand 247667, India

Y. Sharma et al. (eds.), *Recent Research Trends in Energy Storage Devices*,
https://doi.org/10.1007/978-981-15-6394-2_5

the carbon oxidation under OER condition limits their commercial realization. Parallelly, various transition metal-based catalysts have come along as suitable candidates to diminish the burden of precious metal usage [1–3].

Interestingly, the current research is focusing on utilizing insertion materials for electrocatalytic applications because of its well-known working mechanism, and their different crystal structures offer robust catalytic performance. In general, replacing precious metal catalysts is not a perfect solution instead of finding a suitable catalyst with uncompromisable performance and bring them into reality will be the right choice to overcome this crisis. Herein, we introduce a new concept where two different electrochemical reactions take place within the same system. In general, the metal–air batteries are getting prominent interest across the scientific community due to its high energy density. Still, the problem associated with this system is the formation of insoluble discharge product (LiO_2, Li_2O_2) for lithium–air batteries (LABs) and (NaO_2, Na_2O_2) in case of sodium–air batteries (NABs) that lead to reduced cycle life (reversibility) [1].

In this work, a feasibility study of the new hybrid rechargeable battery hinge on two different electrochemical reactions which are OH^- intercalation and Na–air batteries are proved with $NaCo_4(PO_4)_3$ as the active electrode material. Because of combining OH^- intercalation and conventional Na–air reactions, this battery achieves both high-power and energy density, delivering more reliable performance to the conventional sodium–air battery. Besides, this battery manifests good rate capability without any voltage and capacity degradations, which makes this battery system suitable for large-scale utilization such as grid energy storage [4, 5].

2 Materials and Methods

The combustion synthesis route was used to prepare $NaCo_4(PO_4)_3$ material. To acquire the $NaCo_4(PO_4)_3$ XRD patterns, PANalytical X'Pert Pro diffractometer with a Cu-Kα target was used. The GSAS program with the EXPGUI front end was used to perform Rietveld refinement [6]. In order to capture the morphology of these phosphates, the field emission scanning electron microscope (FE-SEM, FEI Inspect F 50 at 5 kV) was used. An FEI Tecnai T 20 U-Twin TEM microscope operating at 200 kV was used for high-resolution TEM of this phosphate sample.

The catalyst ink was made and drop cast on the glassy carbon disk electrodes. Cyclic voltammetry (CV), linear sweep voltammetry (LSV), and long-term durability tests were carried out using 0.1 M KOH with a CH instruments bipotentiostat. For hybrid–air battery cathode material was made with porous $NaCo_4(PO_4)_3$ as an active material, Super-P and polyvinylidene fluoride (PVDF) in the ratio of 80:10:10 weight percent and was coated on the carbon paper. NASICON-type $Na_3Zr_2Si_2P_3O_{12}$ ceramic pellet was employed as the solid electrolyte. The pouch cell assembly was performed in the glove box under high purity Ar gas. Na metal used as an anode was attached with Ni mesh of 9 mm^2 in the glove box. The current collector of the anode side was Ni tab. The final assembled cell had the configuration of (−) Ni mesh |

metallic Na | organic electrolyte | NASICON | 0.1 M NaOH | $NaCo_4(PO_4)_3$ | Ti mesh (+).

3 Results and Discussion

Phosphate $NaCo_4(PO_4)_3$ has been synthesized using the solution combustion synthesis (SCS) technique and is an effective, low-cost process used for producing different nanomaterials which are of use in industrial applications. This method encompasses a self-sustaining exothermic reaction of different oxidizers (e.g. metals) and fuel (e.g. urea, glycine, hydrazides) in a homogeneous solution. It generates extreme exothermic heat that raises the local temperature as high as 600–900 °C within a short duration, which leads to the completion of reaction within the limited duration. The gaseous species breakdown large clusters and develop porosity between particles. Since it is a high-temperature process, so a thermodynamically stable product is formed.

Rietveld refinement of XRD patterns and its structural representation of $NaCo_4(PO_4)_3$ is shown in Fig. 1a, b. The structural parameters obtained were: a = 6.36933 Å, b = 9.8761 Å, c = 15.4704 Å, unit cell volume = 972.71(8) Å^3

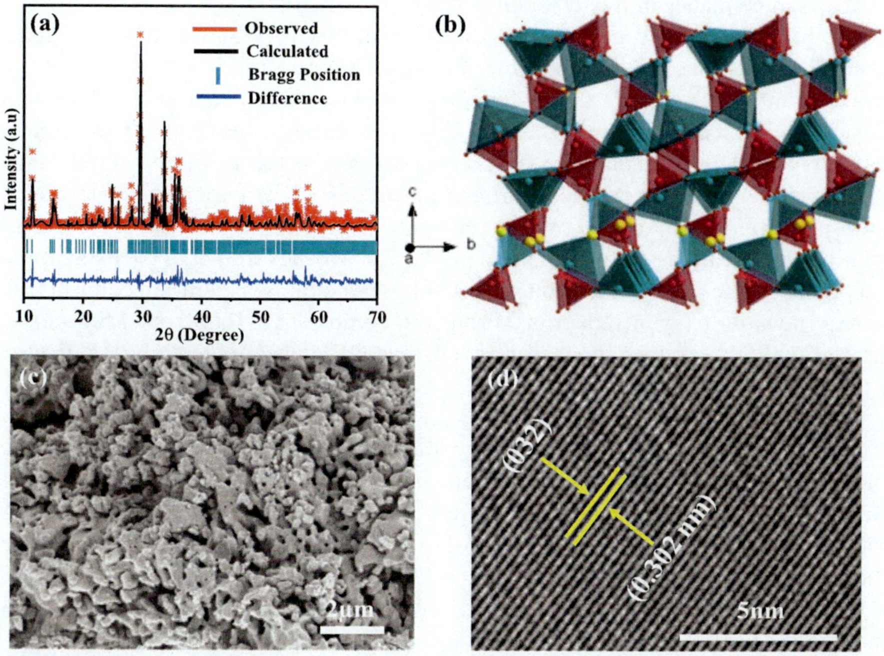

Fig. 1 **a** Rietveld refinement of $NaCo_4(PO_4)_3$, **b** and its crystal structure, **c** SEM morphology, **d** High-resolution TEM image

for $NaCo_4(PO_4)_3$ and crystallizing into monoclinic (*P2/n)* structure. The three-dimensional framework of $NaCo_4(PO_4)_3$ comprises of CoO_5 pyramids and PO_4 tetrahedra, building-large six-sided tunnels along *b* and four and five-sided tunnels that pass along *a* direction. The perspective structure shown is slightly tilted along *a* direction reveals that two Co(1) and Co(2), among the four different types of cobalt polyhedra, consist of CoO_5 tetragonal pyramids which shares their edges and form [CoO_2] ribbons running along *a* direction. The two following ribbons are connected in the direction *b* by a double unit called "Co_2O_9". The latter consists of a tetragonal pyramid, Co(3), and a pyramid Co(4) that is strongly distorted. CoO_5 pyramids therefore form two corrugated [Co_2O_6] layers that are linked in parallel with the (001), linked along *c* through PO_4 tetrahedra [P(1), P(2)] whereas additional PO_4 [P(3)] and further PO_4 tetrahedra inside the layers providing structural stability [7].

The morphology was captured with scanning electron microscopy (SEM) as shown in Fig. 1c. It reveals uniformly distributed porous morphology. Materials with porous structure boost up excellent electrocatalytic activity because of more active site exposure to the guest molecules and allow the electrolyte to penetrate, thereby provides proper wettability that enables fast transport of reactants and electrons.

Inhomogeneity of the pores leads to higher overpotential because of the remote active site to the reactants. However, solution combustion offers uniform distribution which enhances the activity of this material. Figure 1d HR-TEM shows highly crystalline nature of $NaCo_4(PO_4)_3$ and the d-spacing was found to be 0.302 nm matching to (032) lattice planes of this crystalline material.

The bi-functional electrocatalytic properties of $NaCo_4(PO_4)_3$ were investigated in alkali electrolytes for the first time. Figure 2a shows the CV curve recorded in full window which reveals the bi-functional activity of this material. ORR takes place in the cathodic direction whereas OER in the anodic direction with prominent cobalt redox. The CV curves of $NaCo_4(PO_4)_3$ were noted in the potential range of 0.3–1.0 V versus reversible hydrogen electrode (RHE) under an O_2-saturated atmosphere (Fig. 2b). Notable oxygen reduction peaks were seen in the O_2-saturated solution with the high positive potential of ~0.77 V which is similar to the benchmark Pt/C catalyst whereas no reduction peak was observed for N_2 saturation confirms the activity towards oxygen reduction. From the comparative 1600 rpm LSV curve, the $NaCo_4(PO_4)_3$ system showed more positive ORR onset potential ~0.8 V with higher cathodic currents (~−5.1 mA cm^{-2}) which are matching with commercial Pt/C system (~0.95 V & ~−5 mA cm^{-2}). The $E_{1/2}$ (half-wave potential) of $NaCo_4(PO_4)_3$ was found to be ~0.79 V, which is comparable to benchmark Pt/C (~0.86 V). The ORR activity of $NaCo_4(PO_4)_3$ may be due to active ($Co^{II/III}$) centres stabilization by the phosphate framework and their unique coordination facilitates oxygen binding with the ($Co^{II/III}$) active sites. Figure 2d shows OER polarization curves of all three phosphate materials among them $NaCo_4(PO_4)_3$ registered higher onset potential (~1.64 V) as well as current density (~15 mA cm^{-2}) which is better than both RuO_2 and $NaCoPO_4$.

Schematic representation of new hybrid sodium–air/OH^--ion battery is shown in Fig. 3a. The anode compartment which is non-aqueous pouch cell contains Na metal on Ni tab as anode and 1 M $NaClO_4$ in PC as anolyte. The cathode part consists of

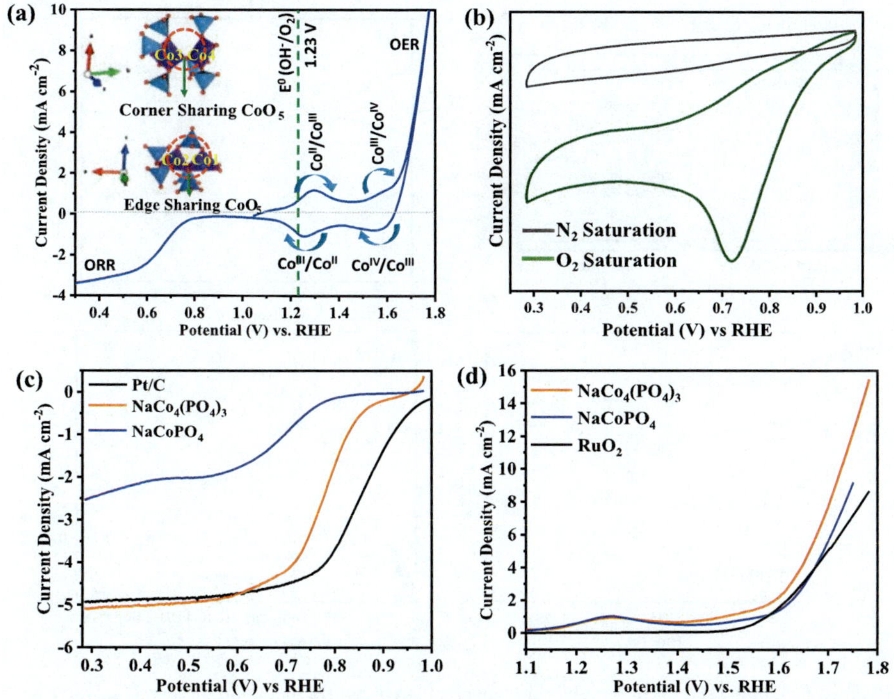

Fig. 2 **a** Bi-functional activity of $NaCo_4(PO_4)_3$ along with cobalt redox, **b** cyclic voltammograms in N_2 and O_2 saturation, **c** a comparison plot of ORR at of 1600 rpm, **d** OER LSV at 1600 rpm

air–cathode and 0.1 M NaOH aqueous electrolyte. These two compartments are kept apart by Na^+ superionic conductor $Na_3Zr_2SiPO_{12}$. The benefits of the hybrid NABs are low overpotential and high-power density, which results from the Na-ion-based solid electrolyte (e.g. $Na_3Zr_2Si_2PO_{12}$) NASICON's high ionic conductivity and easy discharge product (NaOH) solubility. The working mechanism of conventional Na–air battery during charge–discharge at the anode and air electrode can be written as follows: [1]

$$\text{Anode: } Na^+ + e^- \leftrightarrow Na_{(S)},\ E^0 = -2.71\ \text{V}$$
$$\text{Air electrode: } O_2 + 2H_2O + 4e^- \leftrightarrow 4OH^-,\ E^0 = +0.40\ \text{V}$$

Oxygen reduction reaction (ORR) and oxygen evolution reaction (OER) are the two principal reactions that take place in conventional hybrid NABs. The critical difference between traditional NABs and hybrid NAB/OH^--ion battery is the use of multifunctional air–cathodes instead of conventional bi-functional (ORR/OER) catalysts as air–cathode usually porous carbon. Here, we have utilized the porous $NaCo_4(PO_4)_3$ as multifunctional air–cathode, which can show intercalation/deintercalation of OH^- ions and bi-functional electrocatalytic activity because

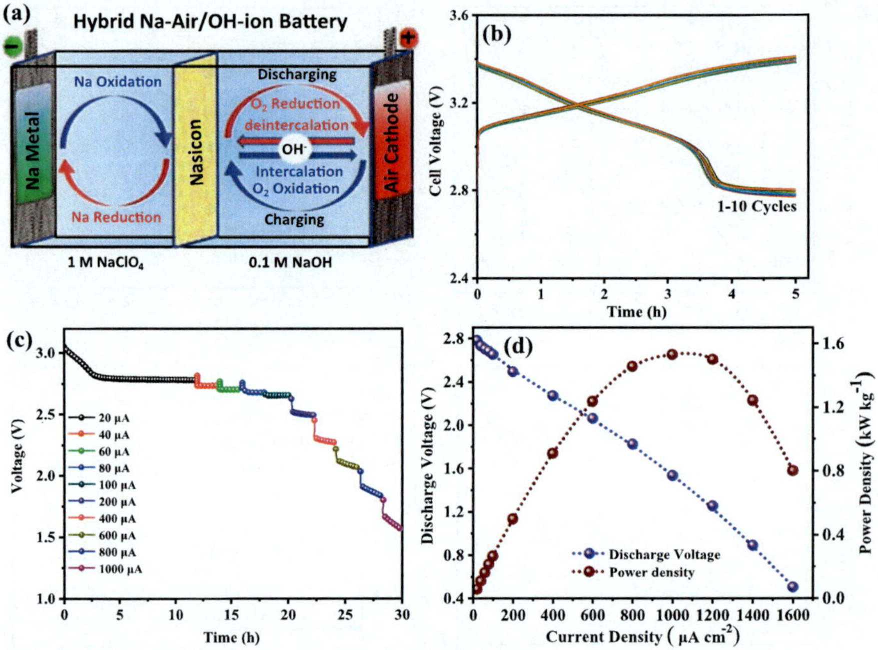

Fig. 3 **a** Schematic representation working mechanism of hybrid metal–air battery, **b** galvanostatic charge and discharge profile at constant current cycled for 5 h, **c** rate capability, **d** power density of $NaCo_4(PO_4)_3$

of its redox activity. Here, the idea is to utilize the well-known (de)intercalation mechanism to store charge before OER/ORR process. The combination of both the mechanisms provides the chance of exploiting the device with high-power and high-energy applications.

The galvanostatic charge–discharge curves of hybrid NABs with the porous $NaCo_4(PO_4)_3$ as multifunctional air–cathode is presented in Fig. 3b. The voltage profiles show both the slope and the stable plateaus. The steady charge and discharge plateau at around 3.6 and 2.8 V correspond to the OER and ORR process which is typical NAB behaviour. The initial charge/discharge process of the NAB up to ~50 mAh g^{-1} is due to (de)intercalation of OH^--ions. The rate capability (Fig. 3c) of this material is excellent. Even at 1000 μA cm^{-2}, it shows stable performance which reveals the stability of these materials. The power density (Fig. 3d) falls at 1000 μA cm^{-2} with a peak power density of 1.5 kW kg^{-1} which is one of the best activities shown by the insertion materials.

4 Conclusion

In summary, the proof-of-concept of a new hybrid battery, which demonstrates electrochemical characteristics of both OH^- intercalation and sodium–air batteries, is demonstrated with the single active material $NaCo_4(PO_4)_3$. Faradaic redox reactions and high specific capacities of the active cobalt species with a flat voltage plateau, resulting in high-power density of this battery system. The new hybrid rechargeable technology is innovative in a range of applications, including electric vehicles, grid energy storage. It provides an opportunity for research and development of integrated power and energy density systems.

Acknowledgements The authors thank the financial support from the Science and Engineering Research Board (SERB, Government of India) in the form of an Early Career Research Award (ECR/2015/000525). B.S. thanks SERB for a National Postdoctoral Fellowship (PDF/2015/00217). The authors thank Shell Technology Centre (STC) Bangalore for financial support.

References

1. Cheon, J.Y., Kim, K., Sa, Y.J., Sahgong, S.H., Hong, Y., Woo, J., Jeong, H.Y., Kim, Y., Joo, S.H.: Graphitic nanoshell/mesoporous carbon nanohybrids as highly efficient and stable bifunctional oxygen electrocatalysts for rechargeable aqueous Na–air batteries. Adv. Energy Mater **6**(7), 1501794–1501803 (2016)
2. Khan, Z., Senthilkumar, B., Park, S.O., Park, S., Yang, J., Lee, J.H., Song, H.-K., Kim, Y., Kwak, S.K., Ko, H.: Carambola-shaped VO_2 nanostructures: a binder free air electrode for an aqueous Na–air battery. J. Mater. Chem., A **5**, 2037–2044 (2017)
3. Yadegari, H., Li, Y., Banis, M.N., Li, X., Wang, B., Sun, Q., Li, R., Sham, T.-K., Cui, X., Sun, X.: On rechargeability and reaction kinetics of sodium–air batteries. Energy Environ. Sci. **7**, 3747 (2014)
4. Manikandan, P., Kishor, K., Han, J., Kim, Y.: Advanced perspective on the synchronized bifunctional activities of P2-type materials to implement an interconnected voltage profile for seawater batteries. J. Mater. Chem. A **6**, 11012–11021 (2018)
5. Lee, D.U., Fu, J., Park, M.G., Liu, H., Kashkooli, A.G., Chen, Z.: Self-assembled NiO/$Ni(OH)_2$ nanoflakes as active material for high-power and high-energy hybrid rechargeable battery. Nano Lett. **16**, 1794–1802 (2016)
6. Toby, B.H.: EXPGUI, a graphical user interface for GSAS. J. Appl. Cryst. **34**, 210–213 (2001)
7. Baies, R., Pe´rez, O., Caignaert, V., Pralong, V., Raveau, B.: A new sodium cobalt phosphate with a tunnel structure, ionic conductor. J. Mater. Chem. **16**, 2434–2438 (2006)

Investigation of Charging and Discharging Characteristics of Lithium-Ion Batteries for Application in Electric Vehicles

Dipanwita Bhattacharjee, Pankaj Kalita, and Sudipta Bijoy Sarmah

Abstract Electric vehicle (EV), which is powered by lithium-ion battery (Li-ion), is an eco-friendly option in today's global warming scenario due to its zero-emission rate. Lithium-ion battery is the most suitable option for an EV owing to its long cycle life, high specific energy, power density, nominal cell voltage, and low self-discharge rate, charging time as compared to other batteries. But, the charge capacity and the charging time of the cell vary with the intensity of the charging current. The charge and discharge characteristics of the battery vary from one manufacturer to another and also between different cells of the same manufacturer. This variation is studied by running tests with constant current and constant voltage in a charge–discharge testing apparatus (Chroma 17011). This study aims to compare the charge and discharge characteristics of two cylindrical cells by subjecting it to a voltage less than its maximum voltage and suggests approaches to reduce undercharging of battery for EV applications.

Keywords Lithium-ion battery (Li-ion) · Electric vehicle (EV) · Constant current (CC) · Constant current constant voltage (CC + CV) · Charge–discharge characteristic

1 Introduction

In a world concerned with the reduction of carbon emissions, the demand for EV is ever increasing. In order to completely replace the gasoline-based vehicles, efficient and reliable EV is the need of the hour. The basic requirement of any good vehicle is that it should run for a long time without interruption. EV is powered by the energy stored in the battery pack. Lithium-ion is the most widely used battery type used in EV due to its specific characteristics and better performance as compared to other batteries. Li- based cells have the highest potential owing to the fact that lithium has the lowest reduction potential as compared to other elements [1]. A large number of

D. Bhattacharjee · P. Kalita (✉) · S. B. Sarmah
Centre for Energy, Indian Institute of Technology, Guwahati, Assam 781039, India
e-mail: pankaj.kalita@gmail.com

Y. Sharma et al. (eds.), *Recent Research Trends in Energy Storage Devices*,
https://doi.org/10.1007/978-981-15-6394-2_6

cells are connected in a string in a battery pack which is used to drive the EV. The individual cells have different characteristic performance owing to manufacturing tolerance, contact resistance, uneven ageing processes etc. Overcharging of cells may lead to explosion and as such protection circuits are provided in a battery pack. Protection circuit stops the charging process as soon as maximum potential is reached by the cells. After prolonged usage, all the cells in a battery pack do not reach its full potential at the same instant during the charging process. The protection circuits cut off the charging process as soon as one of the cells in the battery string reaches its maximum voltage in order to protect it from overcharging. But other cells may not be fully-charged at that instant. These repeated undercharging of the remaining cells results in a reduction of its total storage capacity [2].

In this study, the performance of Li-ion cells when charged up to a voltage lower than its maximum voltage are done. Two cells, which have been previously subjected to 500 charging and discharging cycles under similar condition, are used for the experiment. This is done to replicate the condition of the cells in a battery pack in EV after long usage. One of the cells is charged to 3.7 V (the nominal voltage) instead of 4.2 V (maximum voltage) while the other is charged up to its full potential (4.2 V) in a charge–discharge testing apparatus (Chroma 17011) and their characteristics are compared. Characteristics of two different cells from the same manufacturer both charged up to 3.7 V are also compared. The difference in stored capacity, discharging voltage profiles are plotted.

The charging time and the amount of charge stored in the battery depend on charging current intensity. Slow charging increases the charging time and the stored charge whereas fast charging decreases the charging time and charge storage. The discharging current depends on the power required by the user. The time of discharge depends on the stored charge, the intensity of changing current, etc. [3]. A lot of studies have already been done with low, medium and high current intensities [3–5]. In this study the analysis of charge and discharge characteristics of a commercial Li-ion battery is performed under C-rate 0.136 to 0.9 C in order to study the effects of undercharging on voltage profile, charging and discharging time, stored charge (charge capacity) and extracted charge (discharge capacity).

2 Experimental Procedure

Performance characteristics of two 18650 rechargeable cylindrical Li-ion cell 3.7 V 2200 mAh were studied in this work. The batteries were charged from 3 to 3.7 V using two protocols: Constant Current (CC) and Constant Current Constant Voltage (CC + CV). Cyclic voltammetry (CV) was performed at various scan rates; after the batteries were subjected to charge and discharge of 500 cycles. CV was performed for scan rates 1, 2.5, 5, 10, 20, 25, 50 mV/s in a potentiostat.

For CC protocol, charging current (I_c) of intensity 1.1 A was applied to charge the batteries to 3.7 V and they were discharged to 3 V under constant discharge current (I_d) of 0.3 and 1.1 A for each charging current. After each charging processes, the battery was kept idle for 15 min to ensure relaxation of battery overvoltage before each discharge [6]. For CV + CC protocol, both the batteries were charged from 3 to 3.7 V at a constant current of 0.5 and 1 A, respectively. For the CV step, the battery was kept at 3.7 V until the current decayed to 5 mA. The circuit was opened for 15 min after charging and CC discharging was done for I_d 0.5A and 1.1 A respectively. Then one of the batteries was charged from 3 V to 4.2 V for I_c 0.5 A and 1 A and discharged at I_d 0.5 A and 1.1 A respectively, following the same protocol as that of the first battery. Here, the battery was charged with constant current until it reached 4.2 V. 4.2 V was maintained until the current decayed to 5 mA. After 15 min the discharge cycle was started and the battery was discharged to 3 V. The battery was kept idle for 15 min. to ensure relaxation before the next cycle was started.

Electrochemical impedance spectroscopy was performed for both the batteries at 50% SOC to determine the change in the internal resistance of the battery before and after cycling.

The voltage profile, charging and discharging time, stored charge and extracted charge for all the cycles were recorded and plotted in origin.

3 Results and Discussion

3.1 Cyclic Voltammetry

Initially, the 18650 cells were cycled for 500 times under similar conditions. Then cyclic voltammetry (CV) was performed for scan rates 1, 2.5, 5, 10, 20, 25, 50 mV/s for battery 1 as shown in Fig. 1. Thereafter it was performed for the same scan rates for battery 2 as shown in Fig. 2. No sharp peaks were observed for the CV profile of the two batteries.

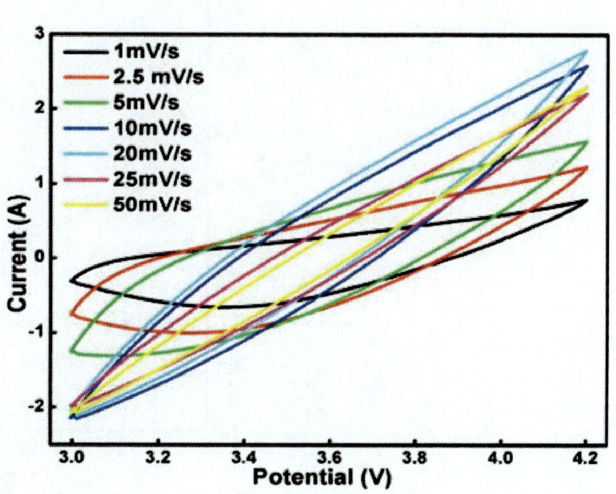

Fig. 1 Cyclic voltammetry profiles with different scan rates for battery 1

Fig. 2 Cyclic voltammetry profiles with different scan rates for battery 2

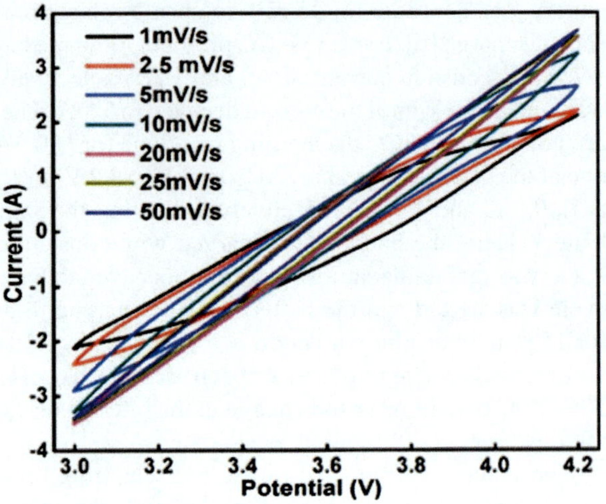

3.2 Constant Current Charging Protocol

Two Li-ion cells were cycled between 3 and 3.7 V under CC I_c 1.1 A and I_d 0.3 and 1.1 A. Their charging, discharging and capacity profiles were plotted.

The voltage and incremental/decreasing capacity profiles of the two batteries for I_c 1100 mA each and I_d 300 mA and 1100 mA, respectively, can be observed in Figs. 3, 4, 5 and 6. Both the batteries were charged from 3 to 3.7 V. Although the batteries were from the same manufacturer, a difference in charging characteristics were observed during undercharging. In both the cases, the first battery took longer time to charge up to 3.7 V thereby increasing the stored charge (capacity) which

Fig. 3 Charge and discharge characteristics under CC charging of 1.1 A and discharging at 0.3 A. Variation of charging and discharging voltage with time

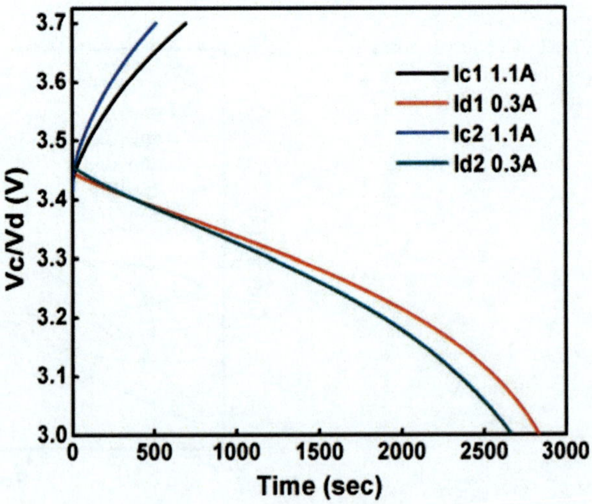

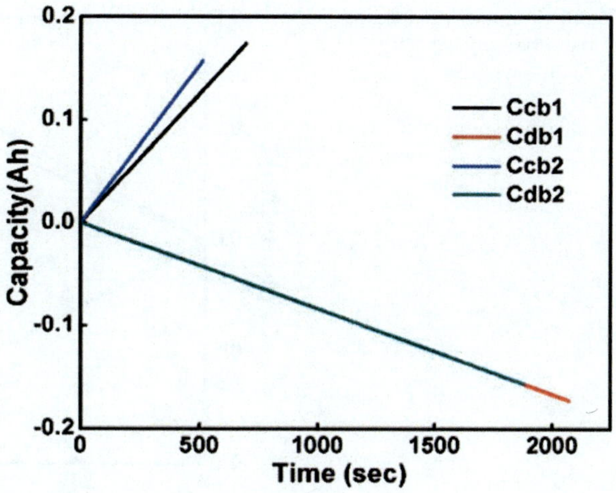

Fig. 4 Variation in charging and discharging capacity with time under CC charging of 1.1 A and discharging at 0.3 A

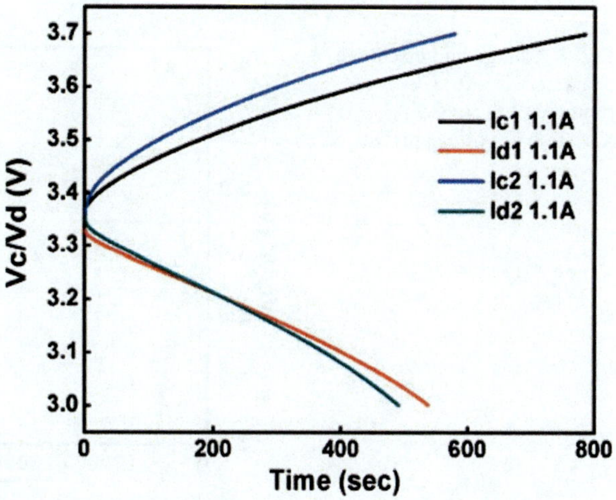

Fig. 5 Charge and discharge characteristics while charging with a constant current of 1.1 A and discharging with 1.1 A. Variation of charging and discharging voltage with time

resulted in longer discharge time even though the same I_d (300, 1100 mA) were applied. The discharge characteristics followed the same trend for both the batteries with a difference in discharge time only. More polarization was observed for higher I_d.

3.3 Constant Current Constant Voltage Charging Protocol

In Fig. 7, 8, 9 and 10 plot of charging voltage profiles for battery 1 charged up to 3.7 V (V_{co1}), battery 2 charged up to 3.7 V (V_{co2}) and battery 2 charged up to 4.2 V (V_{cn}) are plotted against time. The discharge voltage profiles are denoted by V_{do1},

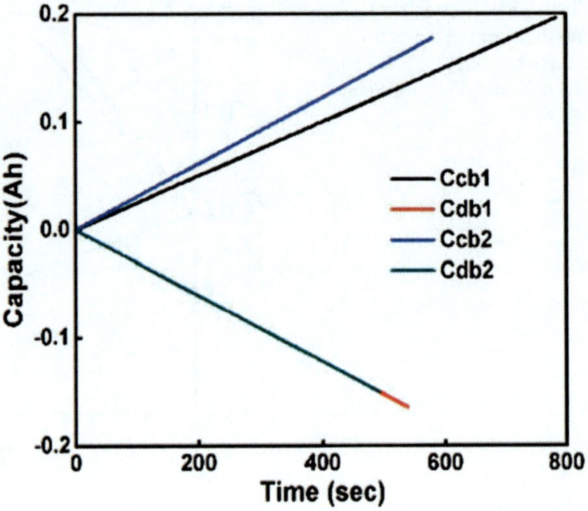

Fig. 6 Variation in charging and discharging capacity with time while charging with a constant current of 1.1 A and discharging with 1.1 A

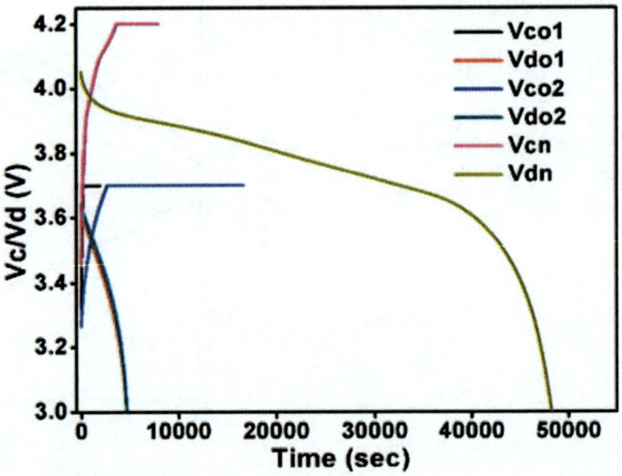

Fig. 7 Charge and discharge characteristics for CC + CV protocol of I_c, I_d 0.5 A. Variation in voltage profile

V_{do2} for battery 1 and 2, respectively, discharged from 3.7 to 3 V, and V_{dn} for battery 2 discharged from 4.2 to 3 V. C_{co1}, C_{co2}, C_{do1}, C_{do2} are the charging and discharging capacity for battery 1 and 2, respectively, when cycled between 3 and 3.7 V and C_{cn}, C_{dn} for battery2 when cycled between 3 and 4.2 V.

There is some variation in charging voltage profiles for the two batteries during undercharging. It is observed that the battery 1 reaches 3.7 V at a shorter time than battery 2. As such, less amount of charge is stored in battery 1 as compared to battery 2. This affects the discharge voltage profile of the batteries. Battery 1 gets discharged a little faster than battery 2. For battery 1, there is a significant increase in discharge time when charged to 4.2 V as compared to 3.7 V. This is because more amount of

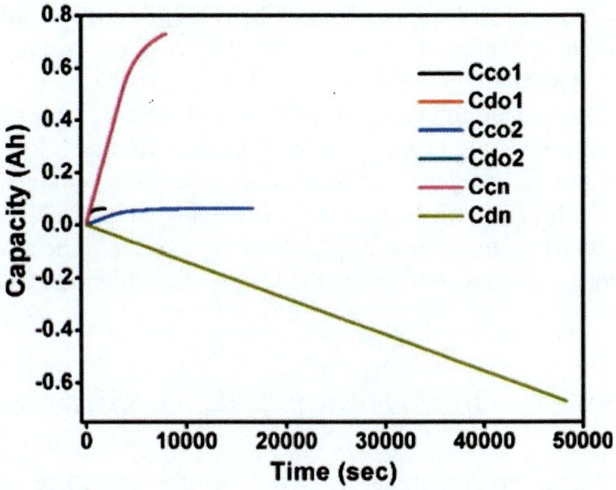

Fig. 8 Variation in charging and discharging capacity for CC + CV protocol of I_c, I_d 0.5 A

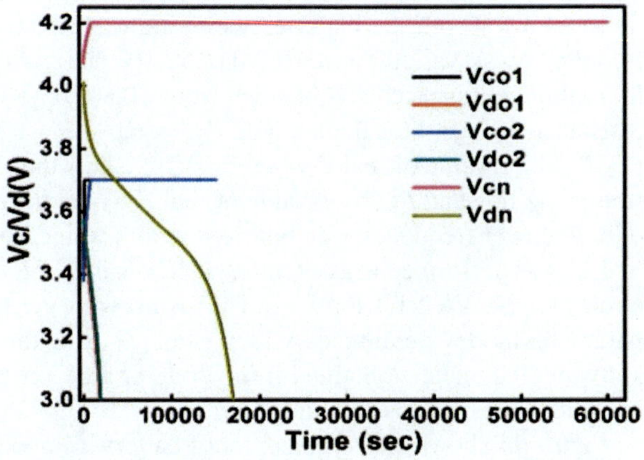

Fig. 9 Charge and discharge characteristics for CC + CV protocol of I_c, I_d 1 A. Variation in voltage profile

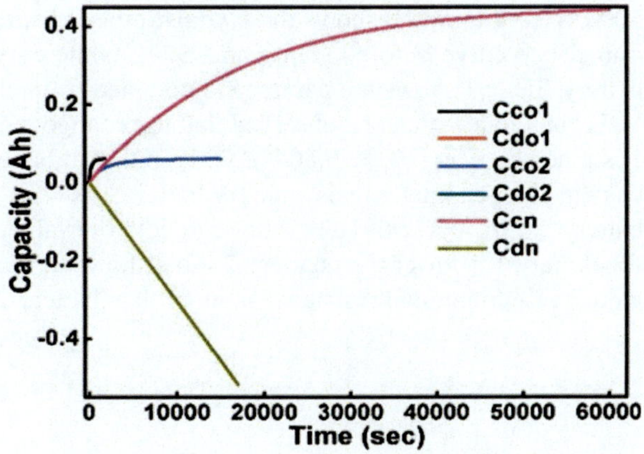

Fig. 10 Variation in charging and discharging capacity for CC + CV protocol of I_c, I_d 1 A

charge is stored in the battery when charged up to its full potential. It can be observed in the variation of capacity with time as shown in Figs. 8 and 10.

The charging capacity is much higher for CC + CV charging up to 4.2 than 3.7 V. The discharging capacity follows a similar trend for both the batteries. C_{do1}, C_{do2} are not visible in Figs. 8 and 10 as they are much lower than C_{dn} and has the same slope. Higher polarization was observed when charged to 4.2 V than that of 3.7 V.

In Figs. 9 and 10, the ohmic resistance of the cell is high and hence the cell did not show the characteristics profile of the material. The discharge capacity of the battery that is displayed is due to the CV mode rather than CC mode.

3.4 Electrochemical Impedance Spectroscopy (EIS)

In order to determine the change in the internal resistance of the battery EIS was performed on the cells before and after cycling. EIS measurement method was performed for both the batteries with sine wave voltage of amplitude 0.01 V. The frequency was varied from 10 MHz to 100 kHz. All spectra are shown with 70 logarithmically spaced frequencies from 100 kHz to 0.01 Hz. The charge transfer resistance R_{ct} can be calculated as $R_{ct} = R_d - R_0$. Here, R_0 is the ohm resistance [7]. R_0 is the value of real Z where the curve cuts the abscissa at $y = 0$ in the higher frequency side and R_d is the value of real Z where the curve cuts the abscissa at $y = 0$ in the lower frequency side before it attains a linearly increasing value.

EIS was performed in a potentiostat. The value of R_d and R_0 were calculated using circle fit in NOVA 2.1.4 software. The results were verified with the equivalent circuit parameter model mentioned by Zhai et al. [8] using the simulation fit in NOVA 2.1.4 software. Parameters obtained from circle fir are Constant phase element (CPE), R_o, R_d. The value of R_{ct} is calculated.

Figure 11 shows the Nyquist plot of battery 1 impedance. B1_1 is the impedance curve before cycling and B1_2 is the curve obtained after cycling the battery for 500 cycles. Figure 12 shows the Nyquist plot of battery 2 impedance. B2_1 is the impedance curve before cycling and B2_2 is the curve obtained after cycling the battery. Table 1 shows the parameters obtained from circle-fit curve fitting.

From Table 1 it can be observed that the change in internal resistance for battery 1 is much higher ($\Delta R_o = 0.80448\ \Omega$) than that of battery 2 ($\Delta R_o = 0.0441\ \Omega$). The change charge transfer resistance for battery 1 is ($\Delta R_{ct} = 1.81464\ \Omega$) is lower than battery 2 ($\Delta R_{ct} = 1.9691\ \Omega$). Although both the values of R_o and R_{ct} before cycling for Battery 1 is lower than battery 2. This shows the degradation of battery 1 due to an increase in internal resistance is more than battery 2.

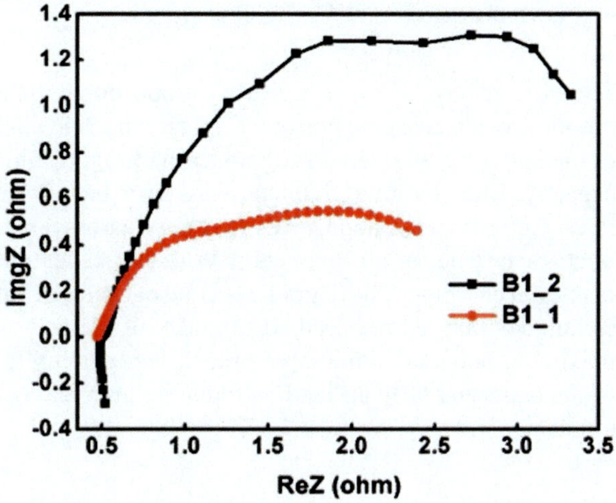

Fig. 11 Plot of battery impedance data curve obtained from EIS with frequency ranging from 100 kHz to 10 MHz. Plot of Imaginary Z Vs Real Z for battery 1

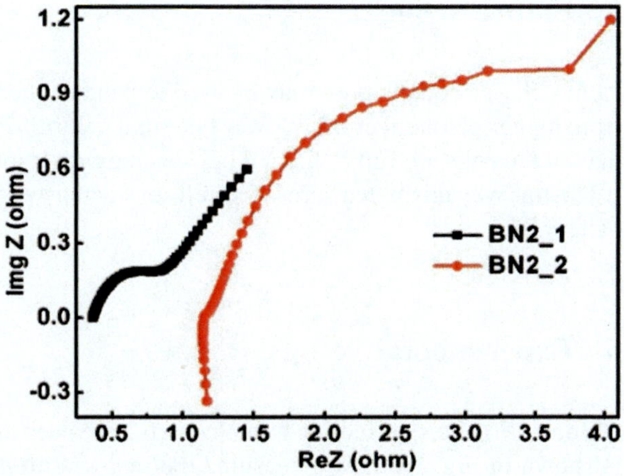

Fig. 12 Plot of battery impedance data curve obtained from EIS with frequency ranging from 100 kHz to 10 MHz. Plot of Imaginary Z Vs Real Z for battery 2

Table 1 Equivalent circuit parameters obtained from circle fit

Obtained/calculated value	Battery 1 before cycling	Battery 1 after cycling	Battery 2 before cycling	Battery 2 after cycling
CPE Y_o (F)	0.19382	1.1706	0.37543	0.49363
R_d (Ω)	0.66838	3.2875	1.3395	3.3407
R_o (Ω)	0.37642	1.1809	0.51247	0.55357
Calculated R_{ct} (Ω)	0.29196	2.1066	0.81803	2.78713

4 Conclusion

The stored charge in a cell is directly proportional to the charging time and inversely proportional to charging current. Capacity much lower than the actual capacity value of the battery is attained during undercharging, resulting in a decrease in its overall capacity. The charging characteristics of a battery depend on previous discharge rates. Current and voltage stress may vary across a battery pack due to the variation of characteristics of different cells. With usage, degradation of cells occurs resulting in uneven charging. The degradation rate of each cell varies with time. As such proper techniques may be required so to make all the cells in a battery string reaches its maximum potential at the same time. It is necessary to ensure that the total capacity of the battery is fully utilized by reducing undercharging in order to use the battery efficiently.

5 Future Scope

Active charge equalization may be used to balance the voltages by extracting energy from higher potential cells to lower potential cells during charging process to ensure that all the cells are fully-charged before charging is cut off by the protection circuit [9]. In this way mismatch across the cells in a battery pack may be reduced, increasing battery life.

6 Experimental Setup

Chroma 17011 testers having four channels have been used to perform the experiment is shown in Fig. 13. DAQ measures the temperature. High-performance industrial computer is used to record data from the battery tester. CV and EIS were performed

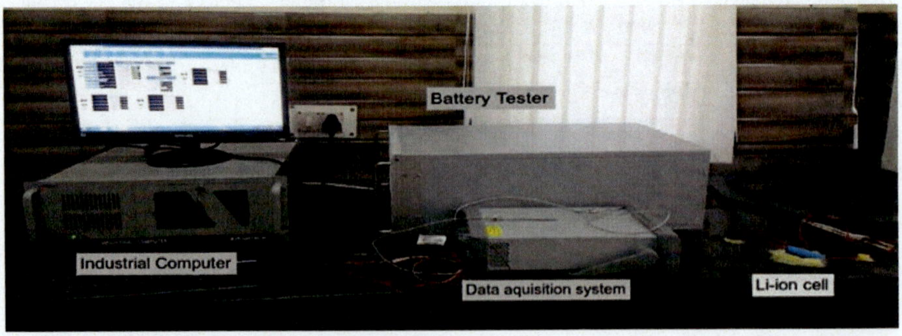

Fig. 13 Chroma 17011 tester with its accessories

in Metrohm Autolab potentiostat N-series and curve fitting was done in NOVA 2.1.4 software.

References

1. Nitta, N., Wu, F., Lee, J.T., Yushin, G.: Li-ion battery materials: present and future. Mater. Today **18**(5), 252–264 (2015)
2. Hannan, M.A., Hoque, M.M., Peng, S.E., Uddin, M.N.: Lithium-ion battery charge equalization algorithm for electric vehicle applications. IEEE Trans. Ind. Appl. **53**(3), 2541–2549 (2017)
3. Da Silva, S.P., Da Silva, P.R.C., Scarminio, A.U.J.: Analysis of a commercial portable lithium-ion battery under low current charge discharge cycles. Quim. Nova **39**(8), 901–905 (2016)
4. Zhang, S.S., Xu, K., Jow, T.R.: Charge and discharge characteristics of a commercial $LiCoO_2$-based 18650 Li-ion battery. J. Power Sources **160**(2), 1349–1354 (2006)
5. Lu, Z., Yu, X., Zhang, L., Meng, X., Wei, L., Jin, L.: Experimental investigation on the charge-discharge performance of the commercial Lithium-ion batteries. In: Yan, J., Chou, S.K., Li, H., Nian, V. (eds.) WES-CUE 2017, vol. 143, pp. 21–26. Energy Procedia, Singapore (2017)
6. Zhang, S.S.: The effect of the charging protocol on the cycle life of a Li-ion battery. J. Power Sources **161**(2), 1385–1391 (2006)
7. Jiang, J., Lin, Z., Ju, Q., Ma, Z., Zheng, C., Wang, Z.: Electrochemical impedance spectra for lithium-ion battery ageing considering the rate of discharge ability. In: Yan, J., Sun, F., Chou, S.K., Desideri, U., Li, H., Campana, P., Xiong, R. (eds.) 8th International Conference on Applied Energy – ICAE 2016, vol. 105, pp. 844–849. Energy Procedia, Beijing (2016)
8. Zhai, N.S., Li, M.W., Wang, W.L., Zhang, D.L., Xu, D.G.: The Application of the EIS in Li-ion batteries measurement. In: 4th International Symposium on Instrumentation Science and Technology, Journal of Physics, Conf. Ser. 48, pp. 1157–1161. IOP Publishing Ltd., China (2006)
9. Shang, Y., Xia, B., Zhang, C., Cui, N., Yang, J., Mi, C.C.: An automatic equalizer based on forward–flyback converter for series-connected battery strings. IEEE Trans. Industr. Electron. **64**(7), 5380–5391 (2017)

Comparative Studies on Crystalline and Amorphous Vinylidene Fluoride Based Fibrous Polymer Electrolytes for Sodium-Ion Batteries

S. Janakiraman, Mohammed Khalifa, Rasmita Biswal, Sudipto Ghosh, S. Anandhan, and A. Venimadhav

Abstract In the present work, electrospun poly (vinylidene fluoride) (PVDF) and poly (vinylidene fluoride-*co* hexafluropropylene) (P(VdF-co-HFP)) fibrous membranes have been compared. Porous homo and copolymer fiber-based membranes with an interconnected structure, high porosity, large electrolyte uptake were prepared by an electrospinning route. The effect of crystallinity in terms of X-ray diffraction (XRD) was investigated for the fibrous polymer membranes (FPMs). The surface morphology of the FPMs is evaluated by field emission scanning electron microscopy (FESEM). The FPMs were soaked in 1 M $NaClO_4$-ethylene carbonate (EC)/diethyl carbonate (DEC) (1:1, vol%) solution to form fibrous polymer electrolytes (FPEs). The ionic conductivity of copolymer showed 1.126 mS cm^{-1} under ambient temperature (at 28 °C) higher than the homopolymer (0.79 mS cm^{-1}) because of HFP unit. The electrochemical stability window of the copolymer membrane also enhanced and stable up to 4.9 V versus Na^+/Na suitable for high voltage sodium rechargeable batteries. When tested with $Na_{0.66}Fe_{0.5}Mn_{0.5}O_2$ as cathode and Na metal as an anode, the cycle performance significantly improved for the copolymer.

Keywords Poly (vinylidene fluoride) · Poly (vinylidene fluoride-*co* hexafluropropyene) · Electrospinning · Ionic conductivity

---

S. Janakiraman (✉) · R. Biswal · S. Ghosh
Metallurgical & Materials Engineering, IIT Kharagpur, Kharagpur, West Bengal 721302, India
e-mail: janakiramiitkgp@gmail.com

M. Khalifa · S. Anandhan
Department of Metallurgical & Materials Engineering, National Institute of Technology Karnataka, Mangaluru, Karnataka 575025, India

A. Venimadhav
Cryogenic Engineering Center, IIT Kharagpur, Kharagpur, West Bengal 721302, India

Y. Sharma et al. (eds.), *Recent Research Trends in Energy Storage Devices*,
https://doi.org/10.1007/978-981-15-6394-2_7

1 Introduction

Sodium-ion batteries (SIBs) are one of the best alternatives to lithium-ion batteries (LIBs) because the risk that the lithium resources cannot cope up with the demands imposed by the automotive and grid storage application and that is the strength of sodium. It is a thousand times more abundant than lithium [1–3]. Compared to the arrangement of basic LIB, a sodium battery also consists of cathode, electrolyte separator system and anode. Among these separator electrolyte systems is a critical component in SIBs that determine the battery's performance, cycling performance and safety. Electrolyte separate system prevents contact between anode and cathode and plays the significant role of transmitting electrons and sodium ions during charging and discharging processes. Gel polymer electrolytes (GPEs) have attractive characteristics over the traditional liquid electrolyte to design safe batteries with lightweight, desired size or shape. Recently, intensive efforts have been made to fabricate a GPE with high ionic conductivity and little interfacial resistance. High dielectric constant and low viscosity are two significant qualities desired of a suitable solvent [4]. GPEs are formed by absorbing a large amount of organic liquid electrolytes in the separator (polymer host) have both the characteristics of solid and the liquid electrolytes, become the favourable choice.

Polymers such as poly(propylene) (PP) [5], poly (ethylene oxide) (PEO) [6], polyacrylonitrile (PAN) [7], poly (vinylidene fluoride) (PVdF) [8], poly(methylmethacrylate) (PMMA) [9] are well examined materials. PVDF and its copolymers are one of the most appealing polymers for separator applications owing due to their high dielectric constant, strong electron withdrawing functional group, thermal properties, and wettability by polar solvents and chemically inert. Various fabrication techniques, such as phase inversion, solvent casting, and electrospinning have been adopted for the improvement for battery separators [10]. Due to the superior properties such as high porosity and large specific surface area, electrospun fibrous membranes can be a good separator for high-performance sodium-ion batteries.

In the present, work electrospun PVDF homo and poly(vinylidene fluoride-co-hexafluoropropylene), P(VDF-HFP) copolymer are compared for their use as battery separators in SIBs. The HFP unit in the copolymer significantly reduces the crystallinity could enhance the ionic conductivity and migration of sodium ions through the amorphous domains. The physical and electrochemical characterization are characterized.

2 Experimentation

2.1 *Fabrication of PVDF and PVdF-Co-hFP Fibrous Polymer Membranes*

The poly (vinylidene difluoride) [PVDF] (Alfa Aesar, average $M_w = 534{,}000$), and poly (vinylidene difluoride-co-hexafluoropropylene) [P(VdF-co-HFP)] (Aldrich, average $M_w = 400{,}000$) and were used as the starting materials for the preparation of fibrous polymer membranes (FPMs) by electrospinning technique. The N, N-Dimethylacetamide (DMAc) and acetone were used as solvents for P(VdF-co-HFP). For the preparation of FPMs, the optimized 16 wt% of P(VdF-co-HFP) polymer solution was dissolved in the mixed solvent of acetone/DMAc (7:3, V/V) under the ambient condition with the help of a magnetic stirrer until a homogeneous solution was obtained. For comparison, PVDF was prepared by electrospinning technique as illustrated elsewhere [11]. The polymer solution for electrospinning is prepared by adding 19% by weight of PVDF to 10 mL of DMSO. The humidity, needle to collector (drum collector) distance, applied voltage, spinning time and the flow rate were 30 ± 2%, 10 cm, 20 kV, 6 h and 0.5 mlph, respectively.

2.2 *Physical and Electrochemical Characterization*

The phase of FPMs was determined by X-ray diffraction (XRD) on a Bruker D8 advance X-ray diffractometer using the Cu Kα radiation (1.5405 Å, 30 mA and 40 kV). The morphology of the FPMs was examined by field emission scanning electron microscopy (FESEM) (Merlin, Germany). The porosities of the FPMs were determined by using n-butanol uptake tests. In an uptake test, the porosity was calculated by using the following equation:

$$\text{Porosity}(\%) = ((M_w - M_d/(\rho_b V)) \times 100 \quad (1)$$

where M_w and M_d are the mass of wet and bare membranes, respectively, ρ_b the density of *n*-butanol, and V is the geometric volume of the FPMs. The electrolyte uptakes were determined by submerging the bare membranes in the liquid electrolyte of 1 M $NaClO_4$ in EC: DEC (1:1, vol%) for 2 h inside the argon-filled globe box at room temperature. The electrolyte uptake was calculated by:

$$\text{Electrolyte uptake, } S_w(\%) = ((W_\text{wet} - W_\text{dry})/W_\text{dry}) \times 100 \quad (2)$$

where W_wet and W_dry are the weights of dry and wet membranes, respectively. The ionic conductivities of the fibrous polymer electrolyte were measured by electrochemical impedance spectroscopy (EIS) using Hioki 3532-50 LCR tester. The impedance measurements were performed on FPEs blocked between stainless steel

(SS) blocking electrodes in a frequency range of 100 kHz to 42 Hz from 28 to 75 °C with an amplitude of 10 mV. The ionic conductivity was determined by:

$$\sigma = t/(R_{\mathrm{FPE}} \times A) \tag{3}$$

where t and R_{FPE} stand for thickness and bulk resistance of the FPE and A is the area of the electrode. The electrochemical stability of the FPEs was determined by linear sweep voltammetry (LSV) using the electrochemical work station. The SS as the working electrode and sodium (Na) foil was used as the counter and reference electrode. The potential window was from 1.0 to 6.5 V at a scan rate of 20 mV s $^{-1}$ at room temperature. The cycling performance of Na/$Na_{0.66}Fe_{0.5}Mn_{0.5}O_2$ cells containing liquid electrolyte-soaked membranes was evaluated using coin-type cells. The $Na_{0.66}Fe_{0.5}Mn_{0.5}O_2$ cathode was prepared by blending active material ($Na_{0.66}Fe_{0.5}Mn_{0.5}O_2$ powder), carbon black conductor and PVDF binder in the weight ratio of 8:1:1, respectively. Arbin battery cycler (BT 2000) was used between 1.5 and 4.0 V. The experiment was done at a current rate of 0.1 C to evaluate the cycling performance.

3 Results and Discussions

3.1 XRD Analysis

Figure 1 displays the XRD profile of PVDF and P(VdF-co-HFP) mats. It can be observed that both the membranes exhibited a crystalline and amorphous region simultaneously. However, by comparing P(VdF-co- HFP) peak at around 20.3° (200) (110) and 36.03° (120) is broader and less intense than compared with the PVDF.

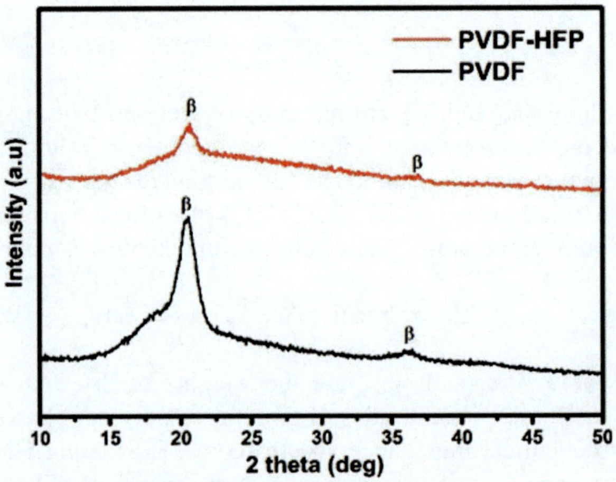

Fig. 1 XRD profile of PVDF and P(VdF-co-HFP) membranes

These observations indicate that the P(VdF-co-HFP) films are of semi-crystalline nature with a substantially large proportion of amorphous regions, which is explained by the presence of the bulky CF_3 groups [12]. Moreover, the amorphous nature of the polymer enhances the ionic diffusivity resulting in greater ionic conductivity of the mats because of higher free volume. The amorphous polymers have a flexible backbone. To further value to the characteristic peaks, PVDF and P(VdF-co-HFP) materials mainly show the stable α—phase crystal structure. The stable characteristics peaks of α–phase is at 18.34° (020), 19.93° (110), 26. 60° (021), respectively. However, the electrospun mats show the diffraction peaks at 20.3° and 36. 03° is belonging to the metastable β phase crystal structure. Probably the stable α—phase would have merged. The conversion from α-phase to β-phase is due to the tensile strain of PVDF jets in a high electric field.

3.2 FESEM Analysis

Morphology and fiber distribution of the PVDF and P(VdF-co-HFP) is shown in the FESEM images (Fig. 2). It can be seen that the electrospun fibrous membranes are smooth and uniformly distributed without any beaded structure. Moreover, the

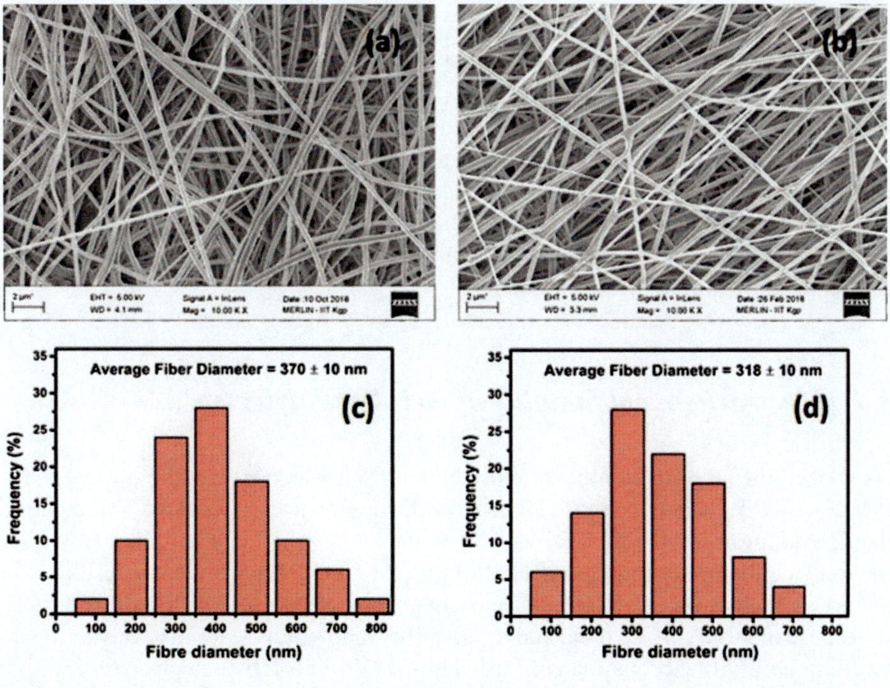

Fig. 2 Typical FESEM images of the **a** and **b** PVDF and P(VdF-co-HFP) mat prepared by electrospinning, **c** and **d** AFD of PVDF and P(VdF-co-HFP) mat

average fiber diameter decreased from 370 ± 10 nm and 318 ± 10 nm. Moreover, porosity of electrospun PVDF and P(VdF-co-HFP) mats measured from n-butanol uptake method is 78 and 83%, respectively. The interlaying fibers produce an interconnected porous structure, which is desirable of the host matrices for polymer electrolytes.

In the electrospinning, a high voltage is used to drive the spinning process. The production of the ultra-fine fibers is obtained by the acceleration and stretching of jets in a high electric field. The population of fibers also increased in P(VdF-co-HFP) mat. At critical voltage, a high charge density is imposed on the ejected jet, thereby the higher stretching forces are introduced to the jet. As a result, the diameter of the fibers progressively reduces with the expansion of applied voltage. It is observed that the numerous pores are responsible for the high conductivity of the sample, i.e., the membrane shows a low average fiber diameter (AFD), and a highly porous structure. This leads to entrapment of large volumes of the liquid in the pores accounting for the increased conductivity in P(VdF-co-HFP). The interconnected micropores in the membrane helped in absorbing liquid electrolytes and hence the ionic conductivity of the membrane is enhanced.

3.3 Porosity and Electrolyte Uptake

The fibrous membranes should have adequate porosity to hold the liquid electrolyte for high ionic conductivity. The porosity of a P(VdF-co-HFP) membrane separator is 83% which is greater than 78% of membrane PVDF. The porosity of the P(VdF-co-HFP) membrane increased with the decreasing fiber diameter can encapsulate more electrolyte within the fibers. The electrolyte uptake capacity of separators affects the performance of Na-ion batteries. The uptake capacity of P(VdF-co-HFP) membrane is 295%, which is naturally higher than that (324%) of PVDF membrane. The high electrolyte uptake of P(VdF-co-HFP) membrane can be explained by its high porosity and reduced fiber diameter value.

3.4 Electrochemical Stability of Gel Electrolytes

As shown in Fig. 3 no significant oxidation current was observed below 4.9 V for P(VdF-co-HFP), whereas for PVDF mat the decomposition starts at 4.5 V, indicating that the addition of HFP in VDF enhanced the electrochemical stability. However, the ionic conductivity mainly occurs through the liquid electrolyte encapsulated in the pores, the contribution to the electrochemical stability of P(VdF-co-HFP) GPE is because of large surface area, and thinner fiber diameter is leading to the improved swelling ability of fibrous matrix [13]. Thus, P(VdF-co-HFP) electrolyte has high anodic stability versus Na/Na^+, and this could be applied to high voltage sodium-ion batteries.

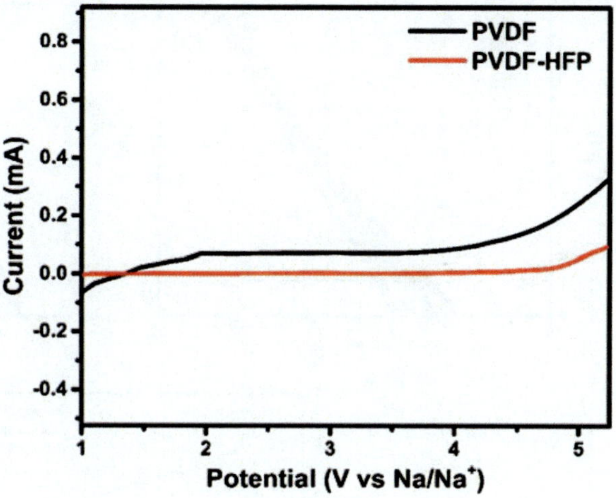

Fig. 3 Linear sweep voltammograms for PVDF and PVDF-HFP fibrous polymer electrolyte after saturating in 1 M $NaCLO_4$ in EC/DEC (1:1 vol%)

3.5 Ac Ionic Conductivity

Figure 4 shows the temperature dependence of ionic conductivity for the PVDF and P(VdF-co-HFP) fibrous polymer electrolytes. It is found that the ionic conductivities for P(VdF-co-HFP) polymer electrolytes (1.12 mS cm^{-1}) are higher than the PVDF (0.79 mS cm^{-1}) in the range of 28–75 °C. The presence of HFP affects the conductivities of the polymer electrolytes over the entire temperature range. The difference in ionic conductivity maybe because of the characteristic molecular level interactions of FPEs. The conductivity mechanism in a gel polymer membrane occurs through the three phases: (1) liquid electrolyte held in the pores (2) amorphous swollen domains and (3) crystalline domains. As a result, the increase in ionic conductivity is mostly due to the liquid electrolyte and swollen amorphous phase in the polymer matrix. Among the three phases, the liquid and the swollen gel phase are the rapid and sluggish conduction paths for ionic conduction. In this context, the existence of HFP increases the amorphous PVDF domain, and decreases the fiber diameter, leading to the high surface area. This aids the fibrous membranes to encapsulate more liquid electrolyte and thus the P(VdF-co-HFP) FPEs exhibits high ionic conductivity. Ionic conductivity, morphology, porosity, mechanical properties and interfacial resistance are considered to be the most influential factors for the growth of dendritic structures on the separator. It is a well-known fact that high ionic conductivity and porous structure are beneficial in the suppression of dendrite growth.

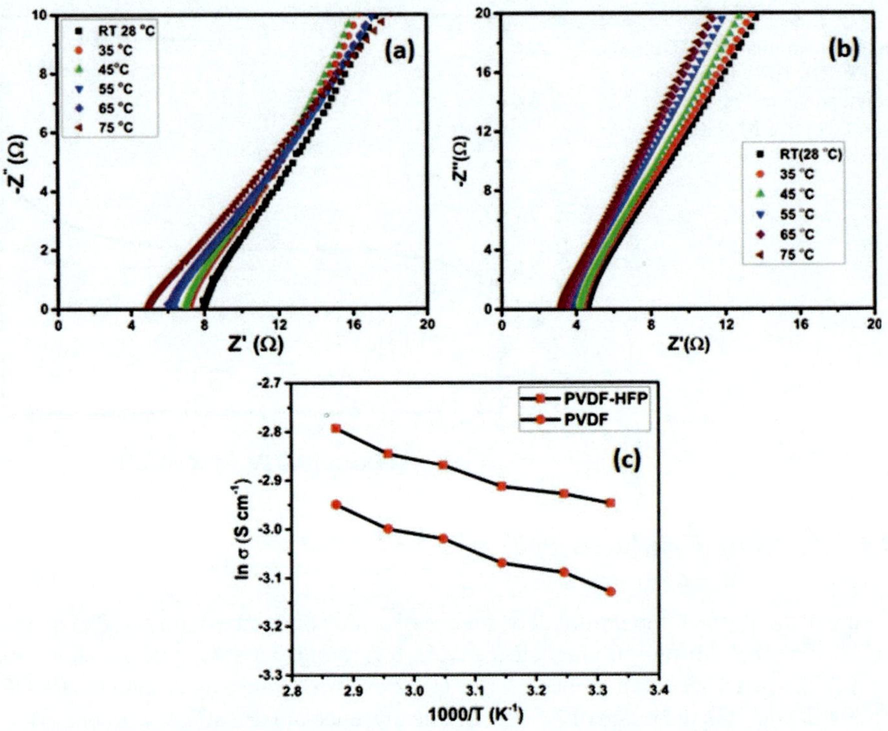

Fig. 4 **a** and **b** Shows the temperature-dependent ionic conductivity of the fibrous polymer electrolyte (PVDF and P(VdF-co-HFP)), **c** comparison of ion conductivities of fibrous polymer electrolyte as a function of temperature

3.6 Second-Cycle Charge-Discharge Curves and Cycling Performance of Cells

Figure 5a shows the second-cycle charge-discharge curves of Na/Na0·66Fe0.5Mn0·5O2 cells containing PVDF and P(VdF-co-HFP) fibrous polymer electrolyte at the 0.1C rate. The second-cycle discharge capacity of the cell is 120 mAh g^{-1} when PVDF membrane is used. The discharge capacity with P(VdF-co-HFP) membrane increased to 123 mAh g^{-1}. The decreased crystallinity and high ionic conductivity of P(VdF-co-HFP) membrane are the main reason for the enhanced capacity values in the cell.

Figure 5b shows the cycling performance of $Na/Na_{0.66}Fe_{0.5}Mn_{0.5}O_2$ at 0.1 C. Interestingly, cell containing PVDF and p(VdF-co-HFP) membrane displayed high specific capacities with a negligible loss after 50 cycles. After 50 cycles, the cell retained a capacity of 79% and 82% of that of the initial cycle for PVDF membrane p(VdF-co-HFP) membrane, respectively. The improved cycling performance of the cell can be attributed to the lower crystallinity and higher ionic conductivity of

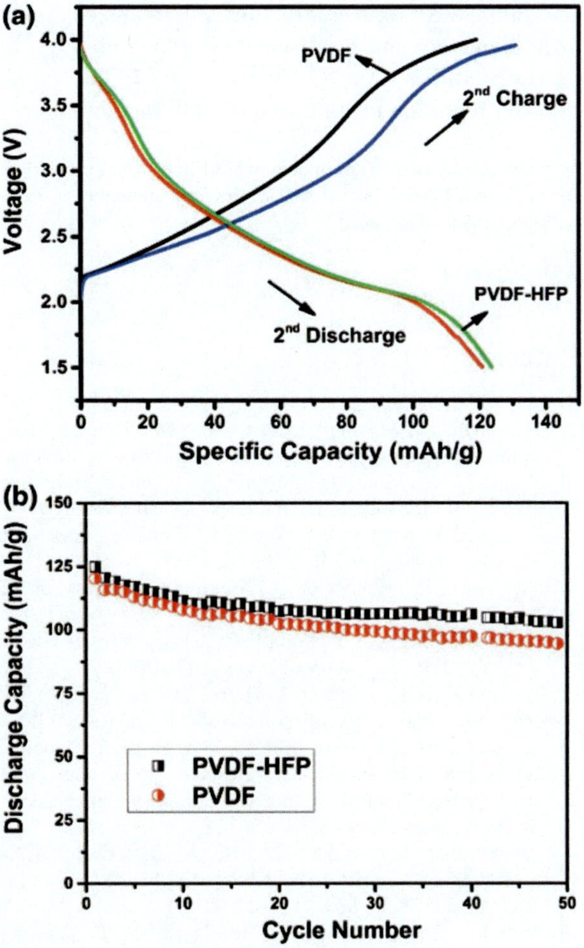

Fig. 5 **a** The second-cycle charge-discharge curves of Na/$Na_{0.66}Fe_{0.5}Mn_{0.5}O_2$ cells containing PVDF and P(VdF-co-HFP) fibrous polymer electrolyte at the 0.1 C rate. **b** Cycling performance of Na/$Na_{0.66}Fe_{0.5}Mn_{0.5}O_2$ cells containing PVDF and P(VdF-co-HFP) fibrous polymer electrolyte at the 0.1 C rate for 50 cycles

the P(VdF-co-HFP) membrane after up taking liquid electrolyte, which could be a promising material for the Na-ion batteries.

4 Conclusions

In conclusion, PVDF and P(VdF-co-HFP) fibrous polymer membranes have been compared for the application of sodium-ion battery. The effect of crystallinity is studied by XRD and FTIR analysis, P(VdF-co-HFP) is comprised of more amorphous domains than the PVDF membrane. The P(VdF-co-HFP) fibrous electrolyte displayed high ionic conductivity of 1.126 mS cm^{-1} under ambient condition because of the HFP unit, providing high conduction paths. The electrochemical stability also improved as a result of fully interconnected network structure, lower average fiber diameters and high specific surface area of fibers. Cycling tests performed on $Na/Na_{0.66}Fe_{0.5}Mn_{0.5}O_2$ half-cells by P(VdF-co-HFP) have revealed very good capacity retention in contrast to PVDF membrane.

Acknowledgements The authors would like to acknowledge the Ministry of Human Resource Development's (MHRD), IMPRINT project grant number 7911 and Ministry of Road and Transport, India for financial support.

References

1. Ponrouch, A., Marchante, E., Courty, M., Tarascon, J.M., Palacín, M.R.: In search of an optimized electrolyte for Na-ion batteries. Energy Environ. Sci. **5**, 8572–8583 (2012)
2. Ponrouch, A., Monti, D., Boschin, A., Steen, B., Johansson, P., Palacín, M.R.: Non-aqueous electrolytes for sodium-ion batteries. J. Mater. Chem. A. **3**, 22–42 (2015)
3. Hwang, J.Y., Myung, S.T., Sun, Y.K.: Sodium-ion batteries: present and future. Chem. Soc. Rev. **46**, 3529–3614 (2017)
4. Stephan, A.M.: Review on gel polymer electrolytes for lithium batteries. Euro. Polym. J. **42**, 21–42 (2006)
5. Song, J.Y., Wang, Y.Y., Wan, C.C.: Review of gel-type polymer electrolytes for lithium-ion batteries. J. Power Sources **77**, 183–197 (1999)
6. Deitzel, J.M., Kleinmeyer, J., Harris, D., Tan, N.C.B.: The effect of processing variables on the morphology of electrospun nanofibers and textiles **42**, 261–272 (2001)
7. Liang, Y., Ji, L., Guo, B., Lin, Z., Yao, Y., Li, Y., Alcoutlabi, M., Qiu, Y., Zhang, X.: Preparation and electrochemical characterization of ionic-conducting lithium lanthanum titanate oxide/polyacrylonitrile submicron composite fiber-based lithium-ion battery separators. J. Power Sources **196**, 436–441 (2011)
8. Janakiraman, S., Surendran, A., Biswal, R., Ghosh, S., Anandhan, S.: Electrochemical characterization of a polar β-phase poly (vinylidene fluoride) gel electrolyte in sodium ion cell. J. Electroanal. Chem. **833**, 411–417 (2019)
9. Rao, M., Geng, X., Liao, Y., Hu, S., Li, W.: Preparation and performance of gel polymer electrolyte based on electrospun polymer membrane and ionic liquid for lithium ion battery. J. Membrane Sci. **399–400**, 37–42 (2012)
10. Lee, H., Yanilmaz, M., Toprakci, O., Fu, K., Zhang, X.: A review of recent developments in membrane separators for rechargeable lithium-ion batteries. Energy Environ. Sci. **7**, 3857–3886 (2014)
11. Janakiraman, S., Surendran, A., Ghosh, S., Anandhan, S., Venimadhav, A.: Electroactive poly(vinylidene fluoride) fluoride separator for sodium ion battery with high coulombic efficiency. Solid State Ionics **292**, 130–135 (2016)
12. Costa, C.M., Silva, M.M., Lanceros-Méndez, S.: Battery separators based on vinylidene fluoride (VDF) polymers and copolymers for lithium ion battery applications. RSC Adv. **3**, 11404 (2013)
13. Wei, W., Dong, C., Tang, Y., Shan, Y., Na, L., Li, Z., Liu, Z.: Shape stability enhancement of PVDF electrospun polymer electrolyte membranes blended with poly (2-acrylamido-2-methylpropanesulfonic acid lithium). Iranian Polym. J. **26**, 179–191 (2017)

Free-Standing Bi-Induced ZrO_2 Nanoflake Array Photoanodes Fabrication for Photoelectrochemical (PEC) Water Splitting Applications

B. Jansi Rani, A. Anusiya, G. Ravi, and R. Yuvakkumar

Abstract Free-standing, self-supporting Bi induced ZrO_2 nanoflake arrays have been synthesized via controlled hydrothermal assisted wet chemical route and have been fabricated as photoanodes by conventional Doctor blade method. Cubic crystal structure of ZrO_2 nanoflake array was confirmed by Raman spectra and by X-ray diffraction study (XRD) predominant peak at 30°. Oxygen and interstitial cation defects oriented with Bi induced ZrO_2 nanoflake arrays were confirmed by photoluminescence (PL) spectra. Deformation stretching mode of Zr–O–Zr vibration confirmed around 492 cm^{-1} through Fourier transform infrared spectra (FTIR). The optical band gap estimation of with and without Bi inducement along with ZrO_2 nanoflake arrays was thoroughly revealed by UV vis diffuse reflection spectra (UV vis DRS) and the calculated band gap values were 3.2, 2.92 and 2.5 eV for 0%, 10% and 20% Bi induced ZrO_2 nanoflake arrays, respectively. Free-standing and self-supporting array formation with clear cut flake edges were examined and confirmed by field emission scanning electron micrograph (FE-SEM) images in 1 μm and 200 nm scale range, respectively. The higher photocurrent density of 19.5 $\mu A/cm^2$ at 1.55 V versus RHE along with smaller Rct associated for 20% Bi induced ZrO_2 nanoflake array photoanode and the excellent photostability over 2 h has been reported for fabricating an efficient PEC cell for solar water splitting applications.

Keywords Free-standing · Bi–ZrO_2 · Nanoflakes · PEC water splitting

1 Introduction

For the past two decades, the goal of researchers is to develop efficient and sustainable energy materials to produce clean energy as an alternative of fossil fuels [1]. Hence, a solar energy source is one of the most conventional resources to produce clean energy through the most convenient and green approach of photoelectrochemical (PEC) water splitting [2]. Several metal oxides are provided a very good outcome to produce

B. Jansi Rani · A. Anusiya · G. Ravi · R. Yuvakkumar (✉)
Nanomaterials Laboratory, Department of Physics, Alagappa University, Karaikudi, Tamil Nadu 630003, India
e-mail: yuvakkumar@gmail.com; yuvakkumarr@alagappauniversity.ac.in

Y. Sharma et al. (eds.), *Recent Research Trends in Energy Storage Devices*,
https://doi.org/10.1007/978-981-15-6394-2_8

clean energy hydrogen via PEC water splitting. Among other metal oxides, ZrO_2 is utilized as efficient due to its wide bandgap, low thermal conductivity, high catalytic stability, etc. [3]. Fabrication of efficient photoanodes for PEC water splitting is still a confront task due to the complexities involved in it such as low carrier lifetime, lack of visible light absorption, and photo-corrosion [4]. Therefore, research activities are attempting to overcome such complexities by introducing surface modifications, bandgap engineering, morphology tuning, dopant incorporation, etc. [5]. Some of the recent research activities achieved the milestone by using ZrO_2 as a potential material. Recently, Li Song and co-workers attempted the surface modification by introducing oxygen vacancies on ZrO_2 supported gold catalyst and subjected the modified electrode for successful water-gas shift reaction applications [6]. Hao and lab mates fabricated dendrite open structure CeO_2–ZrO_2 nanocomposite electrode for water splitting applications [7]. Gaikwad and his research team succeeded in the PEC performance of surface-modified ZrO_2 nanostructures by sulfide species [8]. In this work, we achieved the morphology modification and bandgap engineering by Bi dopant via the optimum hydrothermal method. Among other elements, Bi is one of the most proficient potential materials for PEC water splitting application which will facilitate good photostability and induced the PEC performance towards visible light absorption [9]. Hence, the fabricated photoanodes are thoroughly studied by conventional three electrodes PEC cells.

2 Materials and Methods

Zirconyl nitrate ($ZrO(NO_3)_2{\cdot}xH_2O$), bismuth nitrate pentahydrate ($Bi(NO_3)_3{\cdot}5H_2O$), nitric acid (HNO3), sodium hydroxide (NaOH), deionized water and ethanol were purchased from Sigma Aldrich and used as received. To synthesis bare and Bi induced ZrO_2 nanoflake arrays, the following ration of reagents were used; [0.1 M $ZrO(NO_3)_2{\cdot}xH_2O$ + 1 M NaOH], [0.09 M $ZrO(NO_3)_2{\cdot}xH_2O$ + 0.01 M $Bi(NO_3)_3{\cdot}5H_2O$ + 1 M NaOH], and [0.08 M $ZrO(NO_3)_2{\cdot}xH_2O$ + 0.02 M $Bi(NO_3)_3{\cdot}5H_2O$ + 1 M NaOH] were dissolved in 100 ml deionized water under constant stirring, respectively. For Bi induced ZrO_2 samples, 5 ml of HNO_3 was added to the above solution for dissolving $Bi(NO_3)_3{\cdot}5H_2O$. After that, the precipitating agent 1 M NaOH was dissolved in 100 ml deionized water and added dropwise into the above precursor solution. The solution was then relocated to Teflon lined 250 ml autoclave and maintained at 180 °C for 24 h and then washed with deionized water and absolute ethanol several times. After drying, the powder was calcinated at 500 °C for 3 h to get its cubic ZrO_2 crystalline phase which was named ZR-0, ZR-1, and ZR-2, respectively.

Structural, morphological, and optical properties of the synthesized nanomaterials were investigated as reported elsewhere. The synthesized ZR-0, ZR-1, and ZR-2 active materials were coated using simple Doctor blade method on ultrasonically cleaned 7Ω resistance FTO plate in 1 × 1 cm active area by the following method; 0.1 g of active material was dissolved in 1 ml of ethanol by 10 min of ultrasonication to get homogeneous slurry which was coated on FTO by simple Doctor blade method and left to dry naturally for 24 h. After that, the active material coated FTO substrate was utilized as working electrode in traditional three-electrode PEC cell where Hg/HgO and Pt wire was used as reference and counter electrodes, respectively, in 1 M KOH solution with the help of solar simulator (SOL-3A; Orion) calibrated with a silicon diode equipped with 100 mW/cm^2 xenon lamp as a light source.

3 Results and Discussion

XRD spectra of free-standing Bi induced ZrO_2 nanoflake array photoanodes is represented in Fig. 1a. The results revealed the formation of typical cubic, face-centered ZrO_2 nanoflakes from the matching peaks diffracted at the angles 30.11, 35.00, 50.25, 59.64 and 73.76° relevant to the crystal planes (111), (200), (220), (311) and (400), respectively and all these peaks are concurrently matched with JCPDS data #89-9069. Further, it is clear that, while incorporating Bi ions into ZrO_2 lattice, the peak corresponds to Bi_2O_3 phase started to appear, and also with the increase of dopant

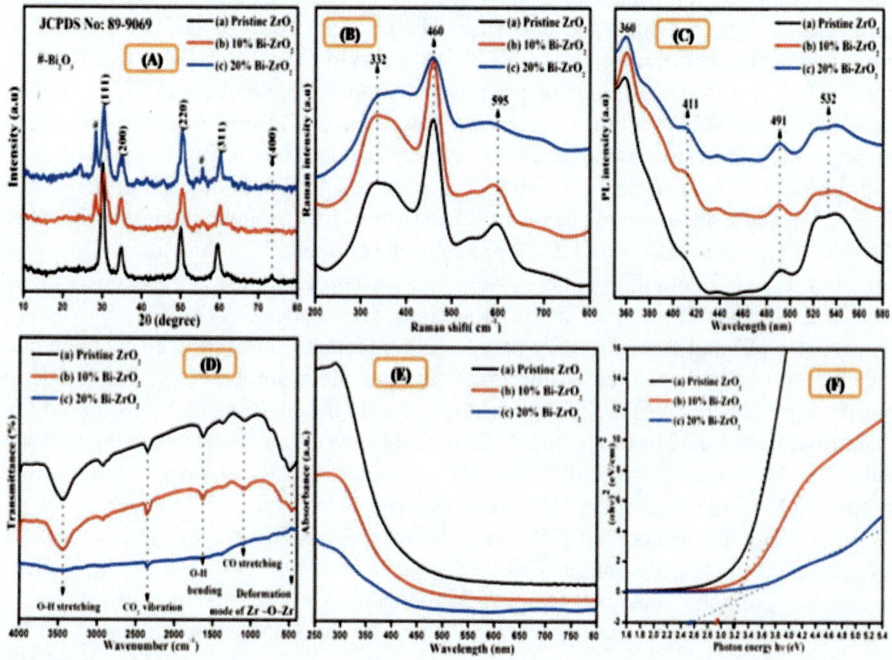

Fig. 1 **a** X-ray diffraction analysis, **b** Raman, **c** photoluminescence spectra, **d** Fourier transform infrared spectra, **e** UV vis diffuse reflection spectra and, **f** Tauc plot of the synthesized bare and Bi induced ZrO_2 nanoflake array

concentration the peak intensity corresponds to Bi_2O_3 phase absolutely increased as shown in Fig. 1a (in # mark) which confirmed the formation of dual-phase occurrence instead of exact replacement of Zr ions by Bi ions. Figure 1b represented the Raman spectra of the synthesized free-standing Bi induced ZrO_2 nanoflake arrays by three strong defined peaks at 332, 460 and 595 cm^{-1} corresponds to the intrinsic phonon characteristics bands of cubic phase ZrO_2 nanostructures as reported in the literature [10]. It is well clear that when introducing and increasing the Bi ions into ZrO_2 lattice, the peak intensity is gradually decreased which may due to the crystal disorder and defect occurrence due to the secondary phase formation of Bi_2O_3 along with ZrO_2 [11].

PL spectra of bare and Bi induced ZrO_2 nanoflake arrays are observed by exciting the sample by 320 nm wavelength. There are four emission peaks observed for all the samples around the wavelength 360, 411, 491, and 532 nm. The peak around 360 nm could be attributed to near band edge emission (NBE) of ZrO_2 nanoflake arrays. The peak at 411 nm could be ascribed to Zr vacancies present in the samples [12]. The other two emission peaks could be attributed to the mid-gap trap states due to the surface defects and oxygen vacancies present in the samples [12]. Further, the reducing intensity of peaks observed for Bi induced ZrO_2 nanostructures compared to bare ZrO_2 stressed out the slower recombination of photogenerated charge carriers which inevitably supported the PEC performance of ZR-1 and ZR-2 samples, respectively. Figure 1d represents the FTIR spectra of the bare and Bi induced ZrO_2 nanoflake arrays in which the absorption bands are observed around 505 cm^{-1} attributed to the deformation modes of Zr–O–Zr bond [13]. The bands around 1095, 1641, 2345, and 3452 cm^{-1} correspond to CO stretching, OH bending, CO2 vibration, and OH stretching vibrations, respectively. Hence, FTIR revealed the formation of ZrO_2 nanostructures and also the intensity decrement due to Bi incorporation.

Figure 1e represents the UV-DRS spectra of ZR-0, ZR-1, and ZR-2 nanoflake arrays. The clear absorbance edge is observed around 400 nm which clearly revealed the redshift arose for ZR-1 and ZR-2 nanostructures due to Bi inducement. Tauc plot shown in Fig. 1f denoted the energy bandgap variation for the synthesized samples as 3.2, 2.92, and 2.5 eV for ZR-0, ZR-1, and ZR-2, respectively. This narrowing mechanism of band structure may be due to the formation of new mid-valence and conduction bands due to the Bi incorporation into ZrO_2 lattice [14]. This bandgap narrowing mechanism of ZrO_2 may strongly due to the increase of VBM due to the interstitial Bi ions and its oxide phases formed the new energy level above the valence band of ZrO_2 which appreciably reduced the bandgap of ZR-2 nanostructures which supported the PEC performance by improved light absorption.

Figure 2a–c represents the FESEM images of ZR-0, ZR-1, and ZR-2 nanoflake arrays at 200 nm scale range. FESEM images undoubtedly portray the typical morphology of nanoflake arrays. Figure 2a represents the petal-like nanoflakes freely standing with some of the agglomerated particle-like nanostructures. Figure 2b represents the modified exact nanoflake array morphology with distorted nanoflake morphology of 10%Bi@ZrO_2 nanostructures. Figure 2c represents the typical free-standing nanoflake array morphology with exact sharp edge hexagonal plate-like flakes with very lesser agglomeration. These images strongly suggest that the

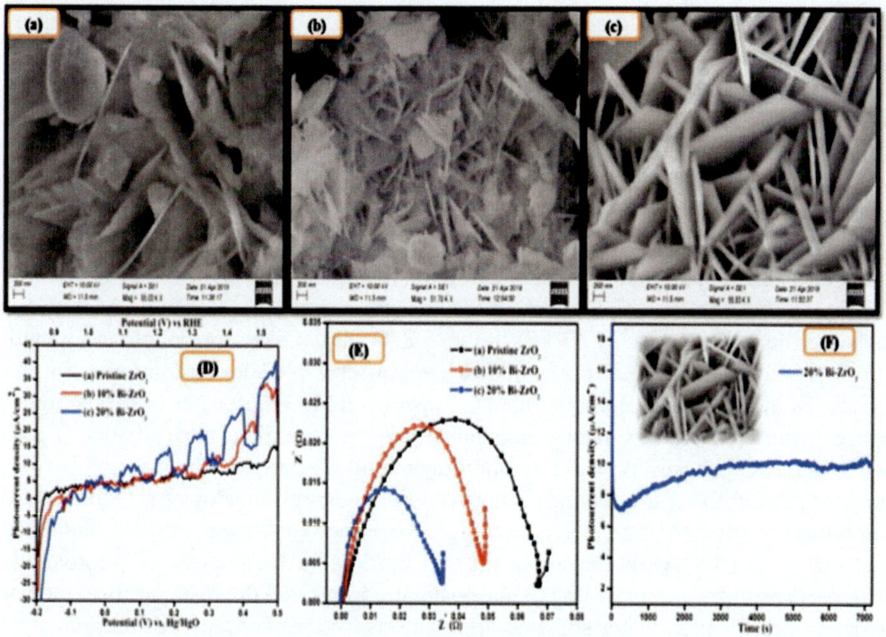

Fig. 2 Field emission scanning electron micrograph images at 200 nm scale range, **a** Bare ZrO_2, **b** 10%Bi@ZrO_2, **c** 20%Bi@ZrO_2 nanoflake arrays, **d** linear sweep voltammogram at 10 mV/s (chopping-condition), **e** electrochemical impedance spectra carried out from the frequency range of 100 Hz to 100 kHz and, **f** Chronoamperometry test for 2 h at 356 mV overpotential of 20% Bi induced ZrO_2 nanoflake array photoanodes

morphology of the ZrO_2 nanoflakes highly tuned by Bi incorporation into it which further improved the formation kinetics and supported the formation of typical free-standing nanoflake arrays with clear cut edges. This kind of morphology improved the conductivity of the photoanode compared to other agglomerated distorted edge morphologies [15]. Hence, the perfect, uniformly distributed free-standing nanoflake arrays supported the overall PEC performance of photoanode under illumination.

Figure 2d represents the LSV graph of fabricated photoanodes under ON-OFF condition at 10 mV/s scan rate. LSV ON-OFF cycle polarization curve clearly denoted the very good response of fabricated photoanodes towards the illumination of light. The actual photocurrent density observed from the polarization curve is 2.6, 11.8, and 19.5 $\mu A/cm^2$ at 1.55 V versus RHE for ZR-0, ZR-1, and ZR-2, respectively. It showed the appreciable enhancement in photocurrent density of four-fold increment for ZR-2 photoanode than bare photoanode. Here, the interesting factor noted that the secondary phase of Bi2O3 formation with ZrO_2 nanoflake arrays during 20% of Bi incorporation strongly supported the PEC response irrespective of the literature reports [11]. Hence, the dual phases of Bi_2O_3 and ZrO_2 synergistically performed well to respond over-illumination to succeed the improved PEC performance. The

PEC mechanism of such dual-phase photoanode might be oriented as nanocomposites which produce the combined effect of enhancement in photocurrent response [16] in addition to the typical morphology of uniform nanoflake arrays. Figure 2e represents the EIS spectra of the fabricated photoanodes that replicated the strong influence of Bi inducement in ZrO_2 lattice. It is clear that the Bi ions introduction into ZrO_2 lattice concurrently reduces the charge transfer resistance (Rct) and further the addition of Bi ions influenced the Rct value and reduced the arc radius thereby improved the charge transfer kinetics under PEC reaction. Figure 2f represents the CA study of the ZR-2 photoanode which exhibited the improved performance in LSV by enhancing the mobility of photogenerated charge carriers. The stability test revealed the superior performance of 20%Bi induced ZrO_2 nanoflake array photoanode with 100% retention in its PEC performance even after 2 h under irradiation of light. Hence, Bi induced ZrO_2 demonstrated the improved PEC performance for its higher concentration of Bi inducement which may due to the synergistic performance of both Bi_2O_3 and ZrO_2 phases, typical free-standing nanoflake arrays with uniform distribution of flakes with clear cut edges, improved light absorption, slower recombination of photogenerated charge carriers and higher conductivity. These results evidenced that the optimized synthetic condition and fabrication techniques of photoanode with perfect uniform morphology synergistically improved the PEC performance of photoanodes for PEC water splitting applications for clean energy production.

4 Conclusions

In summary, an optimized hydrothermal assisted wet chemical route was controlled to yield free-standing, self-supporting nanoflake array morphology of Bi induced ZrO_2 photoanodes. Uniform clear edge morphology of nanoflakes confirmed by FESEM images and the dual-phase nature and the cubic structure of the synthesized material is evidenced by XRD, Raman and FTIR spectra. Improved light absorption supported by the band structure of Bi induced ZrO_2 nanoflake arrays is evidenced by PL, UV, and Tauc plot. Improved PEC performance with higher conductivity and superior stability of 20% Bi induced ZrO_2 photoanode has been demonstrated by LSV, EIS, and CA studies. Hence, the optimized preparation and fabrication of Bi induced ZrO_2 photoanodes obviously utilized as a suitable candidate for solar water splitting clean energy production applications.

Acknowledgements This work was supported by UGC Start-Up Research Grant No. F.30-326/2016 (BSR). This work was supported by UGC-SAP, DST FIST, PURSE and RUSA grants.

References

1. Hernández, S., Gionco, C., Husak, T., Castellino, M., Muñoz-Tabares, J.A., Tolod, K.R., Giamello, E., Paganini, M.C., Russo, N.: Insights into the sunlight-driven water oxidation by Ce and Er-doped ZrO_2. Front. Chem. **6**, 368 (2018)
2. Ellabban, O., Abu-Rub, H., Blaabjerg, F.: Renewable energy resources: current status, future prospects and their enabling technology. Renew. Sustain. Energy Rev. **39**, 748–764 (2014)
3. Reddy, C.V., Reddy, I.N., Shim, J., Kim, D., Yoo, K.: Synthesis and structural, optical, photocatalytic, and electrochemical properties of undoped and yttrium-doped tetragonal ZrO_2 nanoparticles. Ceram. Int. **44**(11), 12329–12339 (2018)
4. Fakhri, A., Behrouz, S., Tyagi, I., Agarwal, S., Gupta, V.K.: Synthesis and characterization of ZrO_2 and carbon-doped ZrO_2 nanoparticles for photocatalytic application. J. Mol. Liq. **216**, 342–346 (2016)
5. Amer, A.W., El-Sayed, M.A., Allam, N.K.: Tuning the photoactivity of Zirconia nanotubes-based photoanodes via ultrathin layers of ZrN: an effective approach toward visible-light water splitting. J. Phys. Chem. C **120**(13), 7025–7032 (2016)
6. Song, L., Cao, X., Li, L.: Engineering stable surface oxygen vacancies on ZrO_2 by hydrogen-etching technology: an efficient support of gold catalysts for water-gas shift reaction. ACS Appl. Mater. Interf. **10**(37), 31249–31259 (2018)
7. Hao, Y., Li, L., Zhang, J., Luo, H., Zhang, X., Chen, E.: Multilayer and open structure of dendritic crosslinked CeO_2–ZrO_2 composite: enhanced photocatalytic degradation and water splitting performance. Int. J. Hydrogen Energy **42**(9), 5916–5929 (2017)
8. Gaikwad, A.P., Betty, C.A., Tyagi, D., Rao, R., Tripathi, A.K., Sasikala, R.: In situ formation of surface sulfide species and its role in enhancing the photocatalytic and photoelectrochemical properties of wide bandgap ZrO_2. Molec. Catal. **435**, 128–134 (2017)
9. Li, H., Bian, Z., Zhu, J., Zhang, D., Li, G., Huo, Y., Li, H., Lu, Y.: Mesoporous titania spheres with tunable chamber structure and enhanced photocatalytic activity. J. Am. Chem. Soc. **129**(27), 8406–8407 (2017)
10. Basahel, S.N., Ali, T.T., Mokhtar, M., Narasimharao, K.: Influence of crystal structure of nanosized ZrO_2 on photocatalytic degradation of methyl orange. Nanoscale Res. Lett. **10**(1), 73 (2015)
11. Rani, B.J., Ravi, G., Yuvakkumar, R., Ravichandran, S., Ameen, F., AlNadhary, S.: Sn doped α-Fe_2O_3 (Sn = 0, 10, 20, 30 wt%) photoanodes for photoelectrochemical water splitting applications. Renew. Energy **133**, 566–574 (2019)
12. Singh, A.K., Viswanath, V., Janu, V.C.: Synthesis, effect of capping agents, structural, optical and photoluminescence properties of ZnO nanoparticles. J. Lumin. **129**(8), 874–878 (2009)
13. Mallick, S., Rana, S., Parida, K.: A facile method for the synthesis of copper modified amine-functionalized mesoporous zirconia and its catalytic evaluation in C–S coupling reaction. Dalton Trans. **40**(36), 9169–9175 (2011)
14. Xiao, M., Li, Y., Lu, Y., Ye, Z.: Synthesis of ZrO_2: Fe nanostructures with visible-light driven H_2 evolution activity. J. Mater. Chem. A **3**(6), 2701–2706 (2015)
15. Rani, B.J., Ravina, M., Ravi, G., Ravichandran, S., Ganesh, V., Yuvakkumar, R.: Synthesis and characterization of hausmannite (Mn_3O_4) nanostructures. Surf. Interf. **11**, 28–36 (2018)
16. Tang, R., Yin, R., Zhou, S., Ge, T., Yuan, Z., Zhang, L., Yin, L.: Layered MoS_2 coupled MOFs-derived dual-phase TiO_2 for enhanced photoelectrochemical performance. J. Mater. Chem. A **5**(10), 4962–4971 (2017)

Silicon/Acetylene Black–Carbon Composite as an Anode for Lithium-Ion Battery

Jay Krishan Dora, Sudipto Ghosh, and Natraj Yedla

Abstract In the current work, the synthesis and electrochemical characterization of silicon/acetylene black–carbon (Si/AB-c) composite has been carried out. Si/AB-c has been synthesized via the combination of sol–gel and the magnesiothermic reduction technique followed by testing as an electrode material in lithium-ion batteries (LIBs). The first discharge and charge capacities for Si/AB-c are impressive and found out to be 2248 and 1914 mAhg^{-1}, respectively. The Si/AB-c composite electrode efficiently accommodates the huge volume expansion encountered during lithiation. Besides, acetylene black enhances electrical conductivity of the electrode material. The versatility to synthesis and design this electrode architecture is quite robust, cost-effective, and reproducible for the large-scale applications in LIBs and many more.

Keywords Composite anode · Magnesiothermic reduction · Lithium-ion battery

1 Introduction

Rechargeable batteries offer widespread usage in portable electronic devices, electric vehicles (EVs), and hybrid electric vehicles (HEVs). Predominantly, lithium-ion batteries (LIBs) attract as a suitable candidate among all the secondary batteries regarding energy density, safety, and cost. Selecting an appropriate electrode material in LIBs has been the major challenge among the research communities [1]. Conventionally, the commercial graphitic anode has limited specific theoretical capacity of 372 mAhg^{-1} [2]. Among the investigated anode materials for LIBs, silicon (Si) holds high theoretical capacity, i.e., ~4200 mAhg^{-1} [3], which delivers ~11 times larger compared to graphite and low discharge potential of ~0.4 V versus Li/Li+,

J. K. Dora (✉) · S. Ghosh
Department of Metallurgical and Materials Engineering, Indian Institute of Technology, Kharagpur 721302, India
e-mail: jay769001.jk@iitkgp.ac.in

N. Yedla
Department of Metallurgical and Materials Engineering, National Institute of Technology Rourkela, Rourkela 769008, India

Y. Sharma et al. (eds.), *Recent Research Trends in Energy Storage Devices*,
https://doi.org/10.1007/978-981-15-6394-2_9

closer to graphite anode [4]. The abundance of Si in nature makes it a cost-effective and eco-friendly electrode. Based on the above facts, silicon and its derivatives are considered as the most favorable anode materials for LIBs. However, some issues prevent the employment of Si as an anode in LIBs. Huge volume expansion (~400%) associated during charge/discharge (lithiation/delithiation) process results in poor cycling life, pulverization of active material, unsafe solid electrolyte interface (SEI), and irreversible capacity loss. Significant attempts have been focused on reducing the size of Si active material to circumvent the aforementioned limitations [5, 6]. Stability of Si has been ensured by implementing numerous forms of nanostructure including nanoparticles, nanowires, nanorods, and composites of Si [7–9]. Synthesis of Si nanostructure involves time-consuming and laboratory limited methods, which restrict it from practical applications. Bao et al. [10] introduced the magnesiothermic reduction technique to overcome the limitations involved in the synthesis of nanostructures. However, this approach has proven to be energy efficient and environmentally feasible for mass production. Several research findings have exhibited superior rate performance and cyclic stability by implementing this technique [11, 12]. This work presents the preparation of silicon/acetylene black–carbon (Si/AB-c) composite from SiO_2/acetylene black carbon (SiO_2/AB-c) via the magnesiothermic reduction route followed by HCl and HF acid etching. The selection of acetylene black carbon as a precursor for the synthesis electrode material is due to its superior conductive behavior, abundance, and economic viability. Furthermore, electrochemical performance of Si/AB-c electrode shows an excellent first cycle discharge and charge capacity. Electrochemical measurements were conducted to evaluate the stability of the electrodes up to 100 cycles.

2 Experimental

2.1 *Preparation of AB-C/SiO_2 Composite*

Initially, AB-c was dried overnight at 120°C in the vacuum. Half a gram of obtained moist-free AB-c was then added in a solution containing 100 mL of ethanol and 50 mL of deionized (D.I.) water and sonicated for 15 min. Subsequently, 5 mL of tetraethyl orthosilicate (TEOS), 3 mL of ammonia solution (28%), and 50 mL of ethanol were added in the solution followed by vigorous stirring for 6 h. The semi-viscous products were centrifuged, washed with ethanol/D.I. water ensuring minimal impurities and then calcined at 500 °C for 2 h in 10% H_2/Ar atmosphere.

2.2 *Preparation of AB-C/Si Composite*

Magnesium powder (Loba Chemie) and calcined AB-c/SiO_2 were mixed thoroughly using a motor pestle. Subsequently, the blend was heated in a tubular furnace at 350°C for 3 h and then at 700°C for 8 h under a 10% H_2/Ar atmosphere and allowed to cool down in the furnace to room temperature. The extracted product was then etched out using 1 M HCl for 8 h followed by HF (5 wt%) treatment for 10 min. Finally, the product was washed and dried overnight in a vacuum at 85 °C.

2.3 *Material Characterization*

Bruker diffractometer was used to record X-ray diffraction (XRD) pattern with Cu Kα radiation ($\lambda = 1.5418$ Å). Field emission scanning electron microscope (FESEM, Melin, Germany) was used to observe the morphologies of the specimen. Raman spectroscopy with argon-ion laser with $\lambda = 514$ nm and 20 MW power was used to observe the effect of Mg and acid on the composite specimen.

2.4 *Electrochemical Measurements*

AB-c/Si composite samples were electrochemically tested using coin cells (CR2032) for lithium-ion cells. The components in the cell include Si/AB-c composite specimen as a cathode, Li foil as an anode, and 1MLiPF_6 in ethylene carbonate (EC): diethyl carbonate (DEC) (1:1 by volume) was used as an electrolyte and polypropylene membrane as a separator. The cell was assembled in an Ar-filled glove box and aged overnight before electrochemical measurements. Galvanostatic charge/discharge cycling was done at a current density of 450 mA g^{-1} in an Arbin automatic battery cycler (BT-2000, USA). The open-circuit voltage was observed as 2.2 V, and the potential window was kept in the range of 0.01 and 1.5 V.

3 Results and Discussion

3.1 *Morphological Studies*

Figure 1a–c shows the X-ray diffraction (XRD) patterns for the SiO_2/AB-c composite; Mg reduced, HCl and HF etched specimen, respectively. Figure 1a shows a broad peak in the range of ~20–30°, which confirms the pattern corresponding to amorphous SiO_2 [13]. Multiple peaks in Fig. 1b infer the successful magnesiothermic reduction of SiO_2/AB-c to Si and MgO. XRD patterns as shown in Fig. 1c assures

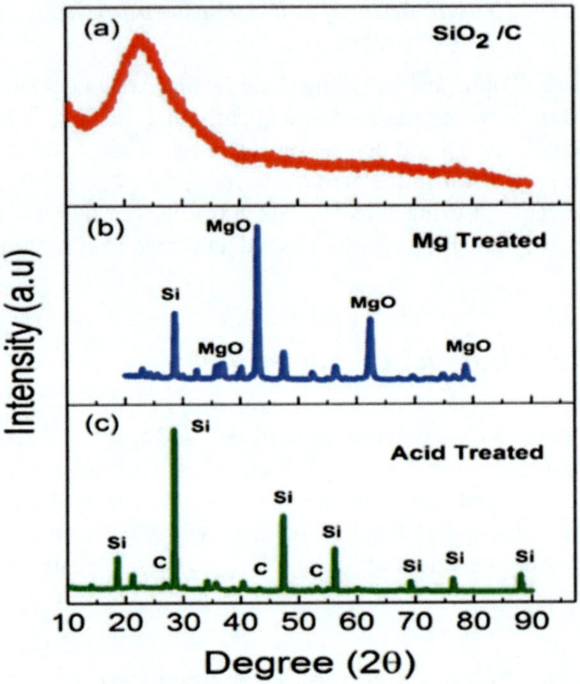

Fig. 1 XRD diffraction patterns of **a** SiO_2/AB-c, **b** Mg-treated, **c** acid-etched specimen

the deduction of unwanted impurities using HCl and HF etching to form crystalline Si–carbon composite.

The evolution of AB-c and Si/AB-c composite structure, as shown in Fig. 3 was also investigated using Raman spectroscopic study. Raman spectra in Figs. 4a and 2b correspond to Si peak at around 513 cm^{-1}, carbon peaks in the range of 1500–1700 cm^{-1} for G band, peak at ~1600 cm^{-1} and 1250–1450 cm^{-1} for D band, and peak at ~1350 cm^{-1}, respectively, [14]. Moreover, a weak broad peak at ~900 cm^{-1} represents the oxide form of Si. Consequently, the spectral peaks of silicon and carbon confirmed the existence of the composite structure.

Figure 3 shows the morphologies of the samples using field-emission scanning electron microscopy (FESEM). The SEM micrographs of Mg-treated SiO_2/AB-c and acid-etched specimen were shown in Fig. 3a, b, respectively. The size of particles was observed to be varying in the range of 130–180 nm. In Fig. 3a, the agglomerated nanoparticles with MgO nanorods were observed after magnesiothermic reduction. MgO was emerged due to the generation of Mg vapor during reduction method through vapor-phase reaction [15]. Figure 3b illustrates the phases of MgO and oxides of Si were dissolved to form spherical Si/AB-c composite particles. Transmission electron microscopy (TEM) image in Fig. 4a further elucidates the internal structure of Si/AB-c composite particles. Energy-dispersive X-ray spectroscopy (EDX) results in Fig. 4b and illustrates the distribution of 'Si', 'C', and 'O'. TEM micrograph in Fig. 4a guarantees the existence of Si and AB-c phases. The EDX peaks observed

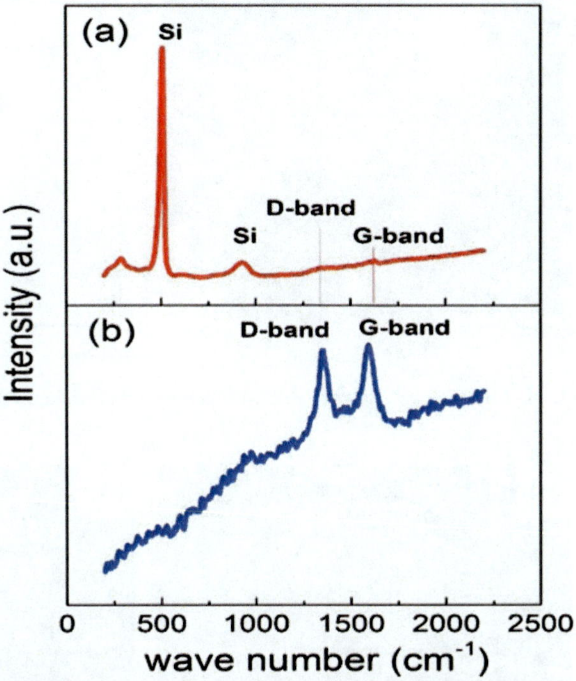

Fig. 2 Raman investigation of **a** Si/AB-c, **b** AB-c

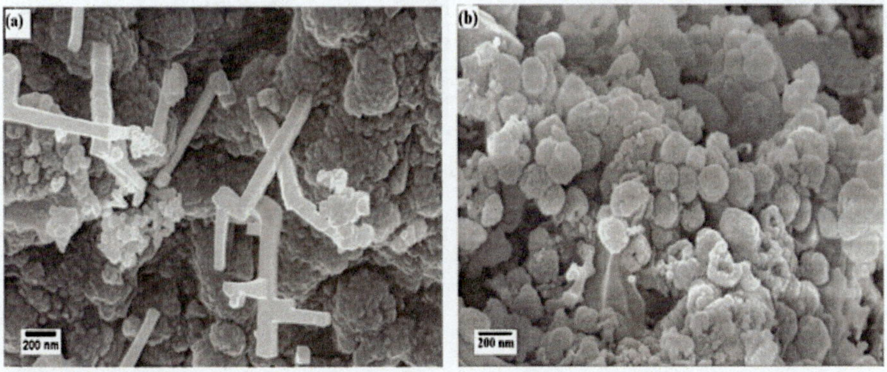

Fig. 3 FESEM image of **a** Mg-treated SiO_2/AB-c, **b** Si/AB-c

in Fig. 4b prove the presence of 'Si', as a strong peak, followed by 'C' and 'O' with lower intensity peaks.

Crystalline silicon nanoparticles with (111) lattice planes were seen to be attached outside the carbon particles, which is depicted in HRTEM image (Fig. 5a) and high-magnification TEM image (Fig. 5b). However, Si particles which sit on the

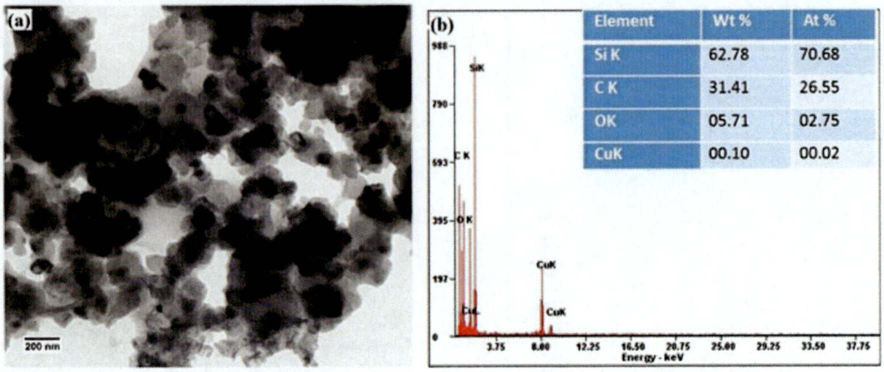

Fig. 4 Morphology analysis, **a** TEM image of Si/AB-c, and **b** EDX mapping of Si/AB-c

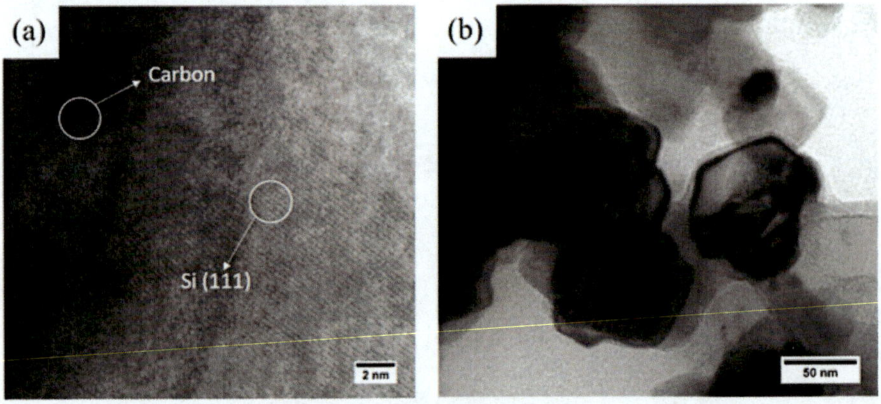

Fig. 5 **a** High-magnification TEM image, **b** high-magnification TEM image of Si/AB-c

surface (shell) deliver high capacity, and carbon core provides mechanical strength and electronic conductivity [17, 18].

3.2 *Electrochemical Studies*

Figure 6a shows the galvanostatic discharge and charge curves of Si/AB-c composite electrode versus Li+/Li cycled between 0.01 and 1.5 V at a current density of 450 mA g^{-1}. The plateau observed at ~0.1 V in the first discharge cycle was due to alloying of Li in the crystalline Si lattice to form amorphous Li_xSi [16]. The discharge (delithiation) and charge (lithiation) capacity in the first cycle for Si/AB-c are 2248 mA h g^{-1} and 1914 mA h g^{-1}, respectively. The corresponding Coulombic efficiency was found to be 85%. Eventually, the irreversible capacity loss was encountered due

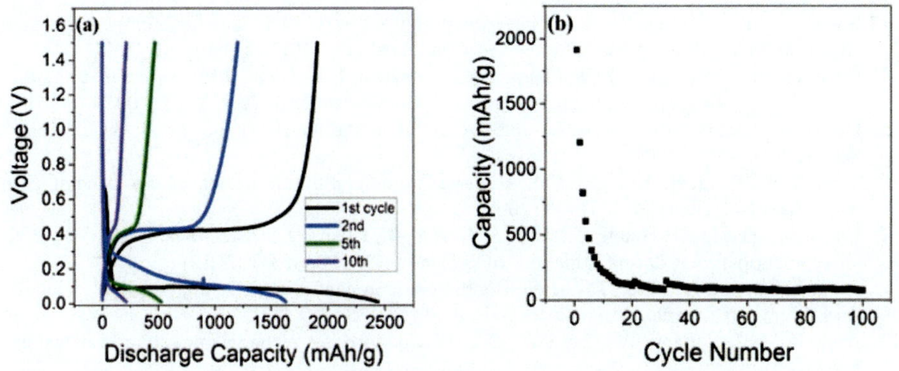

Fig. 6 Electrochemical performance analysis, **a** charge/discharge curve at the current density of 450 mA g^{-1}, and **b** cycling behavior of Si/AB-c

to the emergence of unsafe solid electrolyte interface (SEI) on the electrode surface. Figure 6b exhibits the discharge performance of AB-c/Si electrode up to 100 cycles. It benefits to evaluate and estimate the long-lasting stability of the electrode.

4 Conclusions

Si/AB-c composite was synthesized using simple steps via the magnesiothermic reduction route followed by acid treatment as an anode material for LIBs. The electrochemical test of anode presented an outstanding first cycle discharge and charge capacity with superior Coulombic efficiency. The discharge (delithiation) and charge (lithiation) capacities in the first cycle for Si/AB-c are 2248 $mAhg^{-1}$ and 1914 $mAhg^{-1}$, respectively. Moreover, the magnesiothermic synthesis we undertook confirms as an energy efficient and reproducible approach applicable in varied domains.

Acknowledgements We thank Ministry of Human Resource Development (MHRD), Government of India and Sponsored Research and Industrial Consultancy (SRIC), IIT Kharagpur for their support to carry out this work.

References

1. Tarascon, J.-M., Armand, M.: Issues and challenges facing rechargeable lithium batteries. In: *Materials for Sustainable Energy*. Co-Published with Macmillan Publishers Ltd., UK, pp. 171–179 (2010)
2. Kasavajjula, U., Wang, C., Appleby, A.J.: Nano- and bulk-silicon-based insertion anodes for lithium-ion secondary cells. J. Power Sources **163**(2), 1003–1039 (2007)

3. Song, J., Zhou, M., Yi, R., et al.: Interpenetrated gel polymer binder for high-performance silicon anodes in lithium-ion batteries. Adv. Funct. Mater. **24**(37), 5904–5910 (2014)
4. Beaulieu, L.Y., Eberman, K.W., Turner, R.L., Krause, L.J., Dahn, J.R.: Colossal reversible volume changes in lithium alloys. Electrochem. Solid-State Lett. **4**(9), A137 (2001)
5. Obrovac, M.N., Krause, L.J.: Reversible cycling of crystalline silicon powder. J. Electrochem. Soc. **154**(2), A103 (2007)
6. Chan, C.K., Peng, H., Liu, G., et al.: High-performance lithium battery anodes using silicon nanowires. Nat. Nano. **3**(1), 31–35 (2008)
7. Liu, X.H., Zhong, L., Huang, S., Mao, S.X., Zhu, T., Huang, J.Y.: Size-dependent fracture of silicon nanoparticles during lithiation. ACS Nano **6**(2), 1522–1531 (2012)
8. Hieu, N.S., Lim, J.C., Lee, J.K.: Free-standing silicon nanorods on a copper foil as an anode for lithium-ion batteries. Microelectron. Eng. **89**(1), 138–140 (2012)
9. Peng, K., Jie, J., Zhang, W., Lee, S.-T.: Silicon nanowires for rechargeable lithium-ion battery anodes. Appl. Phys. Lett. **93**(3), 33105 (2008)
10. Bao, Z., Weatherspoon, M.R., Shian, S., et al.: Chemical reduction of three-dimensional silica micro-assemblies into microporous silicon replicas. Nature **446**, 172 (2007)
11. Yu, Y., Gu, L., Zhu, C., Tsukimoto, S., van Aken, P.A., Maier, J.: Reversible storage of lithium in silver-coated three-dimensional macroporous silicon. Adv. Mater. **22**(20), 2247–2250 (2010)
12. Xie, J., Wang, G., Huo, Y., Zhang, S., Cao, G., Zhao, X.: Nanostructured silicon spheres prepared by a controllable magnesiothermic reduction as an anode for lithium-ion batteries. Electrochim. Acta **135**, 94–100 (2014)
13. Wang, H., Wu, P., Shi, H., et al.: Porous Si spheres encapsulated in carbon shells with enhanced anodic performance in lithium-ion batteries. Mater. Res. Bull. **55**, 71–77 (2014)
14. Nemanich, R.J., Solin, S.A.: First- and second-order Raman scattering from finite-size crystals of graphite. Phys. Rev. B. **20**(2), 392–401 (1979)
15. Kim, W.S., Choi, J., Hong, S.H.: Meso-porous silicon-coated carbon nanotube as an anode for lithium-ion battery. Nano Res. **9**(7), 2174–2181 (2016)
16. Li, H., Huang, X., Chen, L., et al.: The crystal structural evolution of nano-Si anode caused by lithium insertion and extraction at room temperature. Solid State Ionics **135**(1), 181–191 (2000)
17. Fan, Y., Zhang, Q., Xiao, Q., et al.: High performance lithium ion battery anodes based on carbon nanotube–silicon core–shell nanowires with controlled morphology. Carbon N Y **59**, 264–269 (2013)
18. Kim, W.S., Choi, J., Hong, S.H.: Meso-porous silicon-coated carbon nanotube as an anode for lithium-ion battery. Nano Res. **9**, 2174–2181 (2016)

Layered $Na_2Mn_3O_7$: A Robust Cathode for Na, K, and Li-Ion Batteries

Krishnakanth Sada, Baskar Senthilkumar, Ritambhara Gond, Valerie Pralong, and Prabeer Barpanda

Abstract Manganese-based layered materials are promising owing to their resource-friendly, low-cost, non-toxic nature with high operational safety. Here, Na–Mn–O ternary layered metal oxide ($Na_2Mn_3O_7$) was synthesized using a single-step conventional solid-state synthesis. The pure phase $Na_2Mn_3O_7$ crystallizes in triclinic structure (s.g. $P-1$). It consists of Mn in 4+ oxidation state with electrochemically active Mn^{4+}/Mn^{3+} redox center. The as-synthesized black powder worked as a robust cathode for the Na-, K-, and Li-ion half-cells. It delivered a reversible capacity of ~140, ~134, and ~160 mA h g^{-1}, respectively, with Na, K, and Li metal as anode in half-cell architecture. Interestingly, Li- ion half-cell exhibited solid-solution-type (de)intercalation upon the variation of Li-ion concentration in $Na_2A_xMn_3O_7$ whereas Na and K-ion (de)intercalation involves two-phase redox reaction.

Keywords Layered oxide · Cathodes · Intercalation

1 Introduction

Growing global concern over pollution and steady usage of rapidly dwindling fossil fuels and non-renewable energy sources has attracted global scientific attention. To address these burning issues, there is an inevitable need to harness renewable energy sources. Renewable energy sources are intermittent, which warrants complementary energy storage devices to store energy upon generation. In this sector, rechargeable batteries play a vital role. The energy stored in the form of chemical potential can be used continuously. From the past three decades, wide range of materials have been studied for battery applications. However, manganese-based materials have a special position with attractive qualities like low cost, environmental-friendly nature,

K. Sada (✉) · B. Senthilkumar · R. Gond · P. Barpanda
Faraday Materials Laboratory, Materials Research Center, Indian Institute of Science, Bengaluru, India
e-mail: skrishnakant@iisc.ac.in

V. Pralong
Normandie Univ, ENSICAEN, UNICAEN, CNRS, CRISMAT, 14000 Caen, France

Y. Sharma et al. (eds.), *Recent Research Trends in Energy Storage Devices*,
https://doi.org/10.1007/978-981-15-6394-2_10

non-toxic, and multiple oxidation states to design high-energy density materials [1]. Accordingly, several Mn-based cathodes like $LiMn_2O_4$ [2] and $LiNi_{1/3}Mn_{1/3}Co_{1/3}O_2$ (NMC) [3] have been commercialized.

The present study demonstrates $Na_2Mn_3O_7$ as a versatile and robust cathode [4]. In the layered $Na_2Mn_3O_7$, Mn is present in the +4 oxidation state and forms layers of MnO_6 octahedra by edge sharing. To examine the suitability of this material, Na-ion was first successfully inserted and extracted from $Na_2Mn_3O_7$ [5–7]. Moving over alternative battery chemistry, K-ion batteries are emerging with added advantages like pre-exist graphite anode, lower standard redox potential (−2.94 V vs. SHE), lower ionic radius (i.e., lower Lewis acidity), and electrode materials with good ionic conductivity. In addition, most of the Li and Na analogue materials are not yet studied for possible application in K-ion batteries. Recently, Sada et al. [8, 9] demonstrated a new concept of using sodium-based materials for K-ion intercalation. As proof of concept, Na_xCoO_2 was successfully employed for K^+ (de)intercalation and it delivered capacity of 82 mA h g^{-1} with end phase of $Na_{0.34}K_{0.5}CoO_2$. This study formed stepping stone to understand the basic structural chemistry as well as to design high-energy density battery materials.

Moving on, after successful demonstration of 'rocking chair mechanism,' Li-ion batteries have become ubiquitous in the portable electronics market. It forms the *numero uno* player in consumer electronics to (hybrid) automobile market. Inspired by the open framework of $Na_2Mn_3O_7$, we investigated its suitability as host for reversible (de)intercalation of Li-ions. The lithiated $Na_2Li_xMn_3O_7$ showed structural transformation from triclinic ($P-$ to trigonal ($R-3m$) framework [4]. The oxide materials are well known for their high capacity, simple synthesis, and structural stability. The practical limit for the maximum reversible capacity of layered $A_2Mn_3O_7$ upon $2e^-$ intercalation/deintercalation is 166, 150, and 180 mA h g^{-1} for Na, K, and Li-ion, respectively, (where A = Na, K and Li). This current piece of work demonstrates layer structured $Na_2Mn_3O_7$ as suitable versatile host for reversible (de)intercalation of Li, Na, and K alkali ions.

2 Experimental

2.1 Materials and Reagents

Sodium nitrate ($NaNO_3$), manganese carbonate ($MnCO_3$), isopropyl alcohol, Super P carbon black, polyvinylidene fluoride (PVDF) binder, N-methyl 2-pyrollidinone (NMP), sodium hexafluorophosphate ($NaPF_6$), potassium hexafluorophosphate (KPF_6), lithium hexafluorophosphate ($LiPF_6$), ethylene carbonate (EC), diethyl carbonate (DEC), propylene carbonate (PC), and dimethyl carbonate (DMC). All above-mentioned chemicals were purchased from Sigma (purity <99%).

2.2 Synthesis of $Na_2Mn_3O_7$

Layered $Na_2Mn_3O_7$ was synthesized by a single-step solid-state synthesis. Initially, stoichiometric amounts of $NaNO_3$ (0.42 g) and $MnCO_3$ (0.86 g) were wet milled using mortar and pestle for nearly 20 min in the presence of acetone. About 5–8 wt% of excess $NaNO_3$ was used to compensate weight loss during high-temperature an nealing process. This precursor mixture was pressed into a dense pellet and was annealed at 600 °C for 4 h in a muffle furnace to get pure phase of the $Na_2Mn_3O_7$. The structure of as-synthesized black powder was analyzed by using high-resolution X-ray diffraction.

3 Results and Discussion

A ternary mixture of Na–Mn–O was synthesized using solid-state synthesis at ambient conditions. The structure of $Na_2Mn_3O_7$ was well characterized using high-resolution X-ray diffraction [8] (PANalytical X'pert pro diffractometer with Cu Kα source $\lambda_1 = 1.5405$ Å and $\lambda_2 = 1.5443$ Å). It crystallizes in highly asymmetric crystal system of triclinic with $P - 1$ space group. The synthesized $Na_2Mn_3O_7$ XRD pattern matches well with PDF# 01-078-0193 [10] pattern (Fig. 1a). Due to the specific orientation in the direction of the (–110) crystallographic plane, all other peaks are suppressed into the background. The inset of Fig. 1a (in the 2θ degree window of 20–80) shows multiple broad peaks. This final product is further characterized using scanning electron microscopy (SEM, Ultra 55 FESEM Carl Zeiss EDS, operated at 5–20 kV) and transmission electron microscopy (TEM, FEI Tecnai F20 S-Twin unit, operated at 200 kV) to know the morphology and the crystallographic orientation of the particles. Figure 1b shows the randomly distributed particles and small agglomerates on the surface of the particles. High-resolution TEM showing the crystallographic planar distance of 5.8 Å and direction of the plane is (010) as shown in Fig. 1c, d. This as-synthesized material was further used for the electrochemical studies without any further modification of the micron-size particles with uniform mixing of Super P carbon black, PVDF, and little amount of NMP as a wetting agent ($Na_2Mn_3O_7$: carbon black: PVDF = 8:1:1). The prepared slurry is uniformly coated on Al-foil and dried overnight in vacuum oven. These electrodes were used to record the electrochemical performance of as-synthesized $Na_2Mn_3O_7$.

Electrochemical properties were studied using the as-prepared electrodes as cathode. The coin cells (CR2032) were assembled in Ar controlled atmosphere using glove box maintained at $O_2 < 0.5$ ppm, $H_2O < 0.5$ ppm. For the systematic optimization of material, starting with Na-ion battery inset of Fig. 2a showing the schematic of the cell assembly, we assembled Na-ion half-cell by using 1 M $NaPF_6$/ethylene carbonate (EC): diethyl carbonate (DEC) (1:1(v/v)) as electrolyte and sodium metal foil as the counter electrode. Further, potassium half-cells were assembled by employing 0.8 M KPF_6 dissolved in EC: DEC (1:1(v/v)) solution as

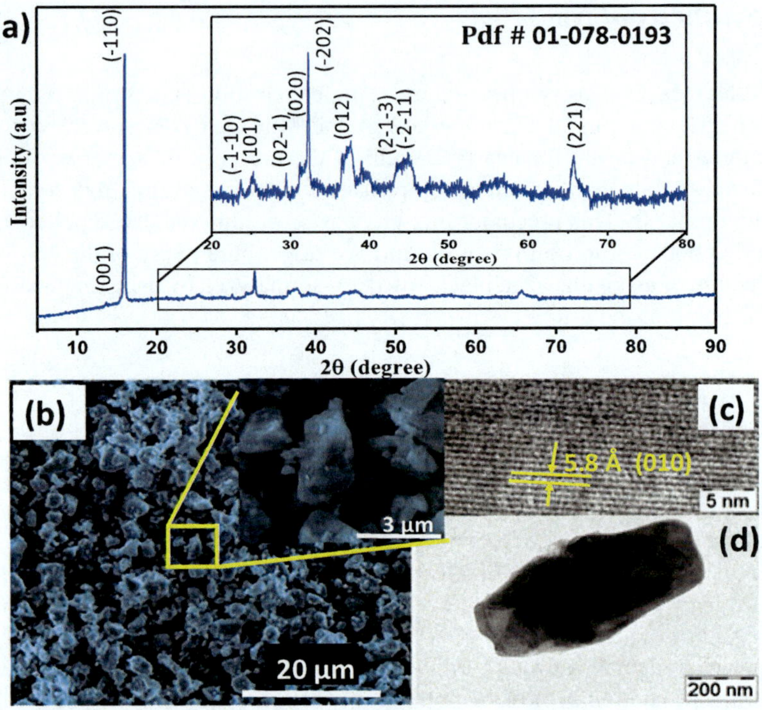

Fig. 1 **a** XRD pattern of the pure phase $Na_2Mn_3O_7$ index with standard pattern (inset showing the multiple peaks of the triclinic crystal system, $2\theta = 20$–$80°$). **b** SEM micrographs with two different magnifications. **c**, **d** HRTEM and TEM images showing specific orientation

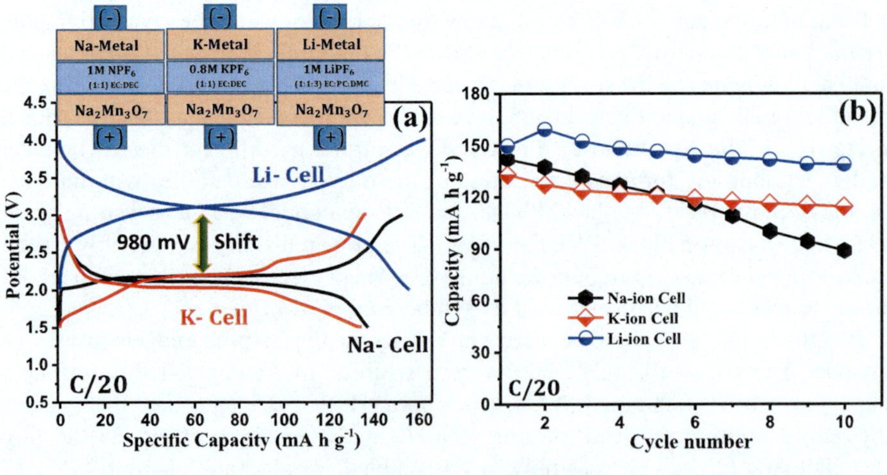

Fig. 2 **a** Comparison of galvanostatic charge and discharge curves of the Na, K, and Li-ion cells. (inset: schematic of the cell assembly with Na, K, and Li metal anode), **b** Cycling stability comparison of the Na, K, and Li-ion cells at the current rate of C/20

electrolyte and potassium metal foil as anode. In similar way, Li-ion half-cells were assembled using 1 M $LiPF_6$ in EC: PC: DMC (1:1:3 v/v %) acting as electrolyte and Li metal foil as counter and reference electrode. In all three cases, glass fiber separator was used by soaking in respective electrolytes. The open framework structure with voids to accommodate the ions in the structure is an advantage to insert any sort of ion as long as it maintains size and the charge of the resulted systems.

Here, transition metal (Tm) is present in oxidized state (Mn^{+4}) so all the studies were carried with initial discharging ($Na_2A_2Mn_3O_7$, A=Na, K, Li). Figure 2a represents the galvanostatic charge/discharge cycles of Na, K, and Li-ion cells with specific capacity of ~140, ~134, and ~160 mA h g^{-1} respectively (Fig. 1a inset showing the schematic of cell assembly). Surprisingly, the voltage upshift between Na, K, and Li is around ~1 V. For the Na and K, average voltage is around ~2.1 V whereas ~3.12 V for the Li-ion half-cell. The possible reasons for this kind behavior one is the standard potential of Li (−3.04 V vs. SHE) and other is structural modification associated with system. The similar kind of voltage variations were observed in the case of Li/$LiCoO_2$ and Na/$NaCoO_2$ cathodes. Also, $Na_2Ti_3O_7$ was studied as anode for both Na (avg. ~0.3 V) and Li-ion (avg. ~1.6 V) half-cells [11, 12]. The galvanostatic cycling stability of the half-cells at rate of C/20 was compared in Fig. 2b. Li-ion half-cell showed better cycling stability compared to the Na and K-ion batteries.

Further, electrochemical properties of the K and Li-ion half-cell were compared in Fig. 3. Firstly, cyclic voltammogram (CV) studies were carried to confine the scanning to the specified electrochemical window to know the redox voltage of the Mn^{4+}/Mn^{3+} couple over the potential of 1–4 V. The redox voltage of the K-ion battery lies between 1.5 and 3 V (Fig. 3a). The galvanostatic charge and discharge were performed between the mentioned potential range. At the rate of C/20, it exhibited

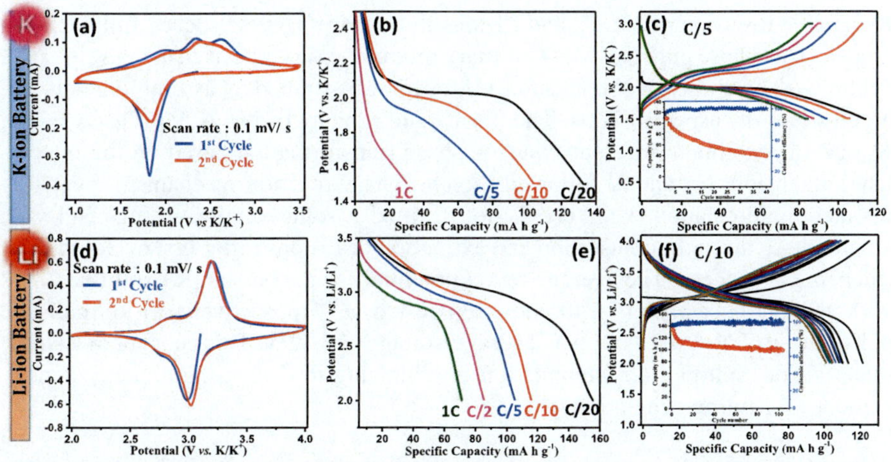

Fig. 3 **a, d** Cyclic voltammogram scan in the range of 1–3.5 V and 2 to 4 V, respectively, for K and Li–ions. (**b, e**). Rate stability studied at different current rates. (**c, d**). Galvanostatic charge/discharge at current rate of C/5 and C/10 for K and Li, respectively

discharge capacity of 152 mA h g^{-1}. The rate stability was tested over different current rates of C/20, C/10, C/5, and 1C that delivered specific capacity values 134, 104, 80, and 32 mA h g^{-1}, respectively, (Fig. 3b). K-ion cell cycling stability was studied at C/5 rate, more than 40 cycles with coulombic efficiency close to 98% as shown in Fig. 3c. The capacity fading was high during low current rate and low during high current rate. It may be due to the larger size of potassium ion and possible Jahn–Teller distortion associated with Mn^{+3} in the discharged state. We can expect the dissolution of Mn into electrolyte due to tetragonal distortion of MnO_6 octahedra. These are the appropriate reasons behind the capacity fading. We can minimize it by coating the electrodes with alumina or substituting with dopants like Ni and Al.

Similarly, Li-ion half-cell architecture was tested within the potential window of 2–4 V. Cyclic voltammogram showed the reduction peak at 3.02 V and oxidation peak at 3.19 V (Fig. 3d). It delivered discharge capacity of 160 mA h g^{-1} at current rate of C/20. Rate stability of the material was studied over the current rates of C/20, C/10, C/5, C/2, and 1C delivering specific capacity values of 160, 116, 106, 86, and 76 mA h g^{-1}, respectively, (Fig. 3e). Cycling stability of system is confirmed at a moderate current rate of C/10, and it retained specific capacity of 105 mA h g^{-1} even after 100 cycles (Fig. 3f). The capacity retention is more than 90%, and inset of Fig. 3f shows the coulombic efficiency to be nearly 100% and cycling stability of Li-ion half-cell over 100 cycles. Li-ion intercalated $Na_2Li_xMn_3O_7$ phase is similar to the $Li_{0.78}Mn_{0.85}O_2$ (#PDF 173134). The ex situ diffraction and PITT studies hinted at single-phase (solid-solution) Li-ion intercalation mechanism [2].

4 Conclusion

Reversible insertion of Na, K, and Li-ions in $Na_2Mn_3O_7$ were successfully demonstrated. This phase pure Na–Mn–O ternary mixture was synthesized using solid-state synthesis. The ability of Na, K, and Li insertion was studied by assembling half-cell architecture of respective ions. The discharged product of $Na_2A_xMn_3O_7$ (A = Na, K, and Li) exhibited different types of phase transformation based on the intercalated alkali ion. Na and K followed the two-phase reaction mechanism reversibly upon (de)intercalation. Unlikely, Li-ion formed a solid-solution mechanism with single-phase throughout insertion and extraction of Li-ions (after first cycle) in electrochemical process. The average redox potential for the Na and K exhibited 2.1 V with specific capacity of ~140 and ~134 mA h g^{-1}, respectively. Li-ion half-cell delivered ~160 mA h g^{-1} with an average voltage of 3.12 V. This concept of a single material for multiple ion intercalation is promising to design high-energy density cathodes for future generation.

Acknowledgements The authors acknowledge financial support from Shell Technology Centre (STC) Bangalore. K.S. sincerely acknowledge Department of Science and Technology (DST, Govt. of India) for financial support under INSPIRE fellowship scheme (IF160154). B.S. thanks DST (SERB), India for providing the National Postdoctoral Fellowship (PDF/2015/00217).

References

1. Thackeray, M., David, W., Bruce, P., Goodenough, J. B.: Lithium insertion into manganese spinels. Mater. Res. Bull. **18**(4), 461–472 (1983)
2. Thackeray, M., Johnson, P. J., de Picciotto, L. A., Bruce, P. G., Goodenough, J. B.: Electrochemical extraction of lithium from $LiMn_2O_4$. Mater. Res. Bull. **19**(2), 179–187 (1984)
3. Johnson, C.S., Li, N., Lefief, C., Vaughey, J.T., Thackeray, M.: Synthesis, characterization and electrochemistry of lithium battery electrodes: $xLi_2MnO_3{\cdot}(1-x)$ $LiMn_{0.333}Ni_{0.333}Co_{0.333}O_2$ $(0 \leq x \leq 0.7)$. Chem. Mater. **20**(19), 6095–6106 (2008)
4. Sada, K., Senthilkumar, B., Barpanda, P.: Layered $Na_2Mn_3O_7$ as a 3.1 V insertion material for Li-Ion batteries. ACS Appl. Energy Mater. **1**(12), 6719–6724 (2018)
5. Adamczyk, E., Pralong, V.: $Na_2Mn_3O_7$: a suitable electrode material for sodium ion batteries? Chem. Mater. **29**(11), 4645–4648 (2017)
6. Sada, K., Senthilkumar, B., Barpanda, P.: Potassium-ion intercalation mechanism in layered $Na_2Mn_3O_7$. ACS Appl. Energy Mater. **1**(10), 5410–5416 (2018)
7. Sada, K., Senthilkumar, B., Barpanda, P.: Potassium intercalation into sodium metal oxide and polyanionic hosts: few case studies. ECS Trans. **85**(13), 207–214 (2018)
8. Sada, K., Senthilkumar, B., Barpanda, P.: Electrochemical potassium-ion intercalation in Na_xCoO_2: a novel cathode material for potassium-ion batteries. Chem. Commun. **53**(61), 8588–8591 (2017)
9. Senthilkumar, B., Sada, K., Barpanda, P.: Layered $P2$-Na_xCoO_2 and Na_xFeO_2 as cathode materials for potassium-ion batteries. ECS Trans. **80**(10), 357–364 (2017)
10. Chang, F., Jansen, M.: Synthesis and crystal structure of $Na_2Mn_3O_7$. Z. Anorg. Allg. Chem. **531**(12), 177–182 (1985)
11. Yabuuchi, N., Kubota, K., Dahbi, M., Komaba, S.: Research development on sodium-ion batteries. Chem. Rev. **114**(23), 11636–11682 (2014)
12. Chiba, K., Kijima, N., Takahashi, Y., Idemoto, Y., Akimoto, J.: Synthesis, structure, and electrochemical Li-ion intercalation properties of $Li_2Ti_3O_7$ with $Na_2Ti_3O_7$-type layered structure. Solid State Ionics **178**(33), 1725–1730 (2008)

Structural and Electrical Transport Investigations on $Mg(ClO_4)_2$-PEO-PEG Solid Polymer Electrolyte Prepared by Hot Press Assisted Route

Munesh Rathore, M. Dinachandra Singh, and Anshuman Dalvi

1 Introduction

The solid polymer electrolytes (SEP), have been extensively studied by ionics community due to various reasons, such as better mechanical properties, ease of fabrication, shape flexibility and promising ability to form proper electrode/electrolyte interface. Several types of Li^+/Na^+ ion-conducting polymer electrolytes have been explored thoroughly. For the last one decade, Mg^{2+} ion-conducting polymers have also got attention due to their several advantages, e.g., low cost, divalent charge, 2:1 anion to cation ratio, high abundance, high safety, ease of handling, low toxicity and the possibility of using more stable magnesium metal as an electrode material [1–7]. Recently, numerous Mg^{2+} ion-based solid polymer electrolytes have been investigated. Recently, R. Mangalam et al. investigated many Mg^{2+} ion-based polymers and found that $Mg(ClO_4)_2$-PVA and $Mg(ClO_4)_2$-PVP solid polymer films exhibit the conductivity in the range of ~10^{-4} Ω^{-1} cm^{-1} at room temperature. These films are predominantly ionic in nature and exhibit the electrochemical stability of 2 V [5].

In another interestingly study, Polu et al. [5] also prepared the Mg^{2+} ion-based polymer films by using PVA and PVP and found the conductivity in the range of 3.44×10^{-5} Ω^{-1} cm^{-1} at 30 °C. They also reported the open-circuit voltage ~1.85 V for the system Mg/[PVA-PVP-$Mg(NO_3)_2$]/(I_2 + C+electrolyte). Hence, Mg^{2+} ion-based solid polymer films appear to offer an interesting option for Mg^{2+} ion solid-state batteries [5].

In the present work, Mg^{2+} ion solid polymer films are synthesized from polyethylene oxide (PEO), polyethylene glycol (PEG), and $Mg(ClO_4)_2$ by solution casting

M. Rathore (✉)
Department of Physics, Amity University Rajasthan, Kant, Rajasthan 303002, India
e-mail: mrchauhan@jpr.amity.edu

M. Dinachandra Singh · A. Dalvi
Department of Physics, Birla Institute of Technology and Science, Pilani, Rajasthan 333031, India

Y. Sharma et al. (eds.), *Recent Research Trends in Energy Storage Devices*,
https://doi.org/10.1007/978-981-15-6394-2_11

followed by hot press assisted route. Preliminary structural, thermal and electrical properties were investigated and reported here.

2 Experimental

The Mg^{2+} ion polymer films of two compositions 10 $Mg(ClO_4)_2$-90PEO and 10 $Mg(ClO_4)_2$-90(0.6PEO-0.4PEG) were successfully prepared solution casting method. The appropriate amount of mixture of polyethylene oxide (PEO), polyethylene glycol (PEG), and $Mg(ClO_4)_2$ were dissolved in acetonitrile using a magnetic stirrer for 3 h. The solution was subsequently taken in a petri dish and dried for 24 h. These films were further treated in a polymer hot press (at ~80 °C, 1 tons) to obtain flat uniform films.

To scrutinize the structural characterizations of these polymer films, X-ray diffraction (Rigaku, MiniFlex II) and Fourier-transform infrared spectroscopy (FTIR) have been performed. The thermal investigations were carried out by using Differential Scanning Calorimetry (DSC). The electrical conductivity measurements (30–70°C) were carried out at a controlled heating rate of 1 °C/min using computer interfaced HIOKI Impedance Analyzer model 3532-50 in the frequency range 42 Hz–5 MHz. The ionic nature was confirmed by dc polarization technique. Linear sweep voltammetry (LSV) was further performed to investigate the electrochemical stability by using Princeton 263A potentiostat/galvanostat on cells of type SS/sample/SS.

3 Results and Discussion

The X-ray diffraction (XRD) patterns for as prepared polymer films for Pure PEO, 10 $Mg(ClO_4)_2$-90PEO, and 10 $Mg(ClO_4)_2$-90(0.6PEO-0.4PEG) are shown in Fig. 1. The appearance of peaks corresponds to PEO are visible in all the samples. However, the intensity of PEO peaks decreases with the addition of PEG and 10 $Mg(ClO_4)_2$. As apparent, PEG completely dissolves in the polymer matrix. These results suggest the composite nature of the samples.

The FTIR spectrum for as prepared polymer films is obtained at room temperature in the range 400–1800 cm^{-1} (Fig. 2). The peak positions are identified with standard stretching in various groups [4, 6–8] as given in Table 1. For Pure PEO, the C–H bending is seen at ~1466 cm^{-1}. It is observed that the position C–H bending as observed in pure PEO shifts towards lower wave number when PEG is added. This apparently suggests bond weakening due to PEG addition. In the case of Mg^{2+} ion contained films, additional ClO_4^- asymmetric bending is observed.

The DSC scans for as prepared films are obtained at heating rate of 10 °C/min and shown in Fig. 3. The endothermic dip is observed in all the samples which suggest the melting of PEO. For Pristine PEO, a broad endothermic peak appears at ~59 °C that may be attributed to melting of PEO. For 10 $Mg(ClO_4)_2$-90PEO, the melting event

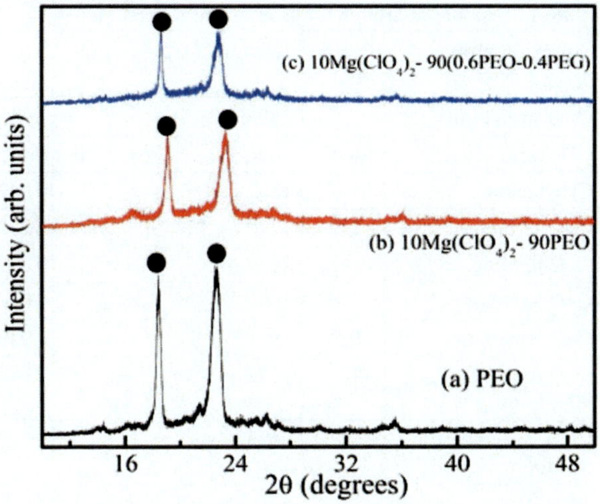

Fig. 1 X-ray Diffraction patterns for different films: **a** PEO, **b** 10 $Mg(ClO_4)_2$-90PEO, **c** 10 $Mg(ClO_4)_2$-90(0.6PEO-0.4PEG)

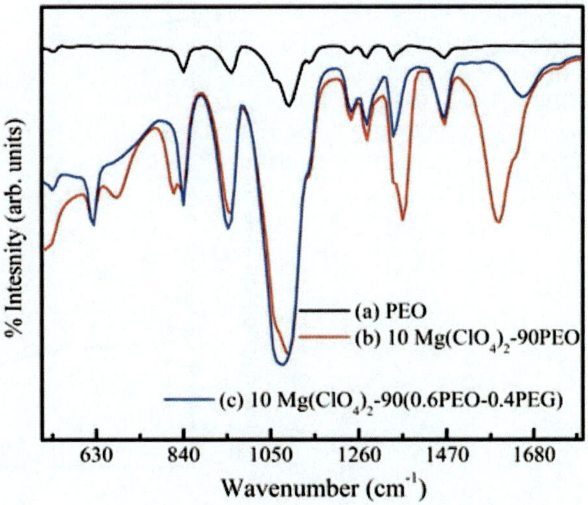

Fig. 2 FTIR spectra for **a** PEO, **b** 10 $Mg(ClO_4)_2$-90PEO, **c** 10 $Mg(ClO_4)_2$-90(0.6PEO-0.4PEG)

onset is apparent at ~40 °C. For 10 $Mg(ClO_4)_2$-90(0.6PEO-0.4PEG), the initiation of melting event observed at ~47 °C. Hence addition on PEG slightly improves the thermal stability of Mg^{2+} ion-based containing polymer films.

The electrical properties, conductivity-temperature cycles are shown in Fig. 4. The electrical conductivity exhibits an Arrhenius like behavior for both the compositions. The results are perfectly reproducible between room temperature to ~343 K. It is found that the addition of PEG leads to improvement in the ionic conductivity. A typical value of $\sim 10^{-5}\ \Omega^{-1}\ cm^{-1}$ is observed at room temperature.

Figure 5 shows the variation of conductivity as a function of frequency for the 10 $Mg(ClO_4)_2$-90PEO and 10 $Mg(ClO_4)_2$-90(0.6PEO-0.4PEG) polymer films at

Table 1 Functional groups for as prepared films

Structural group	Pure PEO	10 $Mg(ClO_4)_2$-90PEO	10 $Mg(ClO_4)_2$-90(0.6PEO-0.4PEG)
C=O stretching	–	1595	1655
CH_2 scissoring	1466	1466	1462
CH_2 bending	1342	1342	1342
C–O–C binding	1279	1279	1279
C–O binding	1243	1243	1243
C–O stretching	1096	1092	1076
C–C stretching	956	954	948
CH2 rocking vibrations	842	842	842
ClO_4^- asymmetric bending	–	623	623

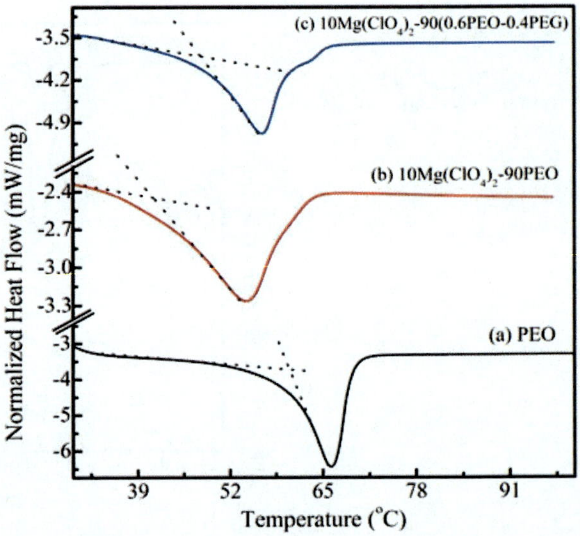

Fig. 3 DSC pattern for **a** PEO, **b** 10 $Mg(ClO_4)_2$-90PEO, **c** 10 $Mg(ClO_4)_2$-90(0.6PEO-0.4PEG). Dotted line indicates onset of melting event

40 °C. In lower frequency region, the conductivity is independent of the frequency with a slight decrease for $\omega \to 0$ due to polarization. Further, at higher frequencies, the conductivity exhibits dispersive nature which is a typical solid polymer electrolyte behavior.

The Nyquist plots obtained for 10 $Mg(ClO_4)_2$-90(0.6PEO-0.4PEG) polymer films are shown in Fig. 6. It plot exhibits depressed semicircle followed by an inclined line that may be modeled as a parallel combination of resistance and constant phase element in series with another constant phase element. The low-frequency inclined line confirms the predominant ionic nature of the samples. The diameter of the

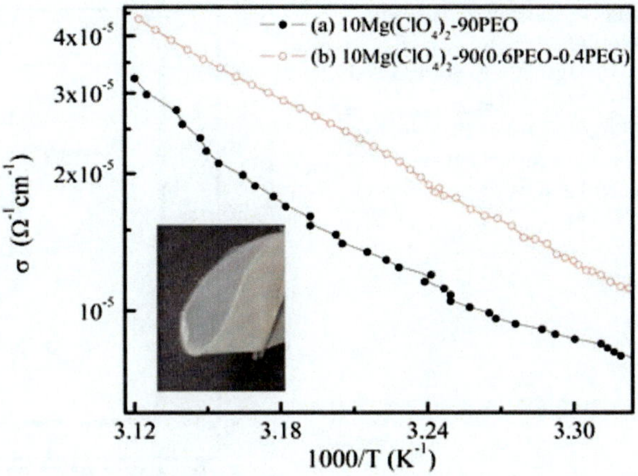

Fig. 4 Electrical conductivity- Temperature cycles for as prepared films

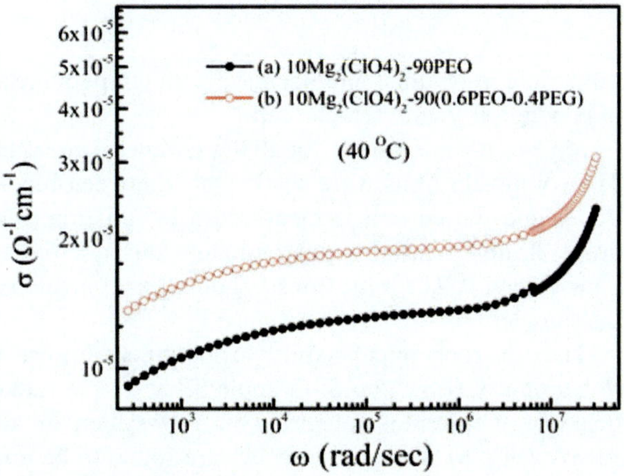

Fig. 5 Electrical conductivity versus frequency for **a** 10 $Mg(ClO_4)_2$-90PEO, **b** 10 $Mg(ClO_4)_2$-90(0.6PEO-0.4PEG)

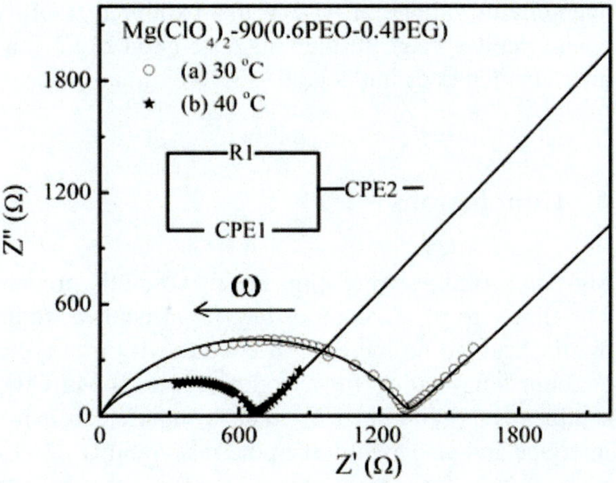

Fig. 6 Nyquist plot for 10 $Mg(ClO_4)_2$-90(0.6PEO-0.4PEG). Equivalent circuit is also shown in the inset

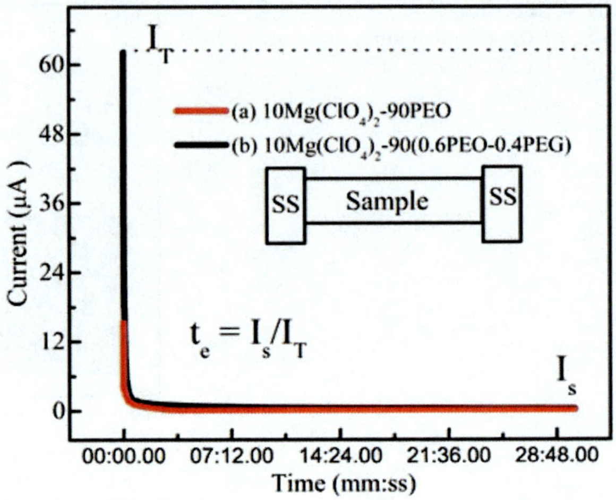

Fig. 7 Current versus time plot for **a** 10 $Mg(ClO_4)_2$-90PEO and (b) 10 $Mg(ClO_4)_2$-90(0.6PEO-0.4PEG). Ionic transport number was calculated using $t_+ = 1 - t_e$

semicircle in Nyquist plot decrease with temperature which confirms the increment of conductivity with temperature.

Ionic transport number for all the samples is measured by dc polarization (Fig. 7). The composite films were sandwiched between blocking stainless steel (SS) electrodes and the current is measured as a function of time. Apparently, the current drops abruptly initially and saturates subsequently which confirms the dominant ionic behavior. The ionic transport number is calculated is found to be ~0.95 for both the samples.

The electrochemical stability of polymer film was measured from Linear Sweep Volatmmetry (LSV) on SS/Sample/SS at a scan rate of 10 mV/s. As apparent, the behavior of current is not abrupt but very systematic that further exhibits stability of ~1.8 V (Fig. 8). These composites are found to be promising for solid-state battery applications. Interestingly, electrochemical cell of the type Mg/SPE/I_2 exhibits an open-circuit voltage of ~1.7 V that exhibited stability for at least 72 h. The stable open-circuit voltage further suggests good Mg^{2+} ion activity in the SPE/electrode interface that does not degrade for long time.

4 Conclusions

Mg^{2+} ion solid polymer films are successfully prepared by hot press assisted route. The films are of the composite type in nature. Addition of PEG into the polymer matrix leads to an increase in conductivity to a typical value of ~10^{-5} Ω^{-1} cm^{-1} at room temperature for a composition 10 $Mg(ClO_4)_2$-90(0.6PEO-0.4PEG). The samples are predominately ionic in nature, do exhibit good Mg^{2+} activity at the interface and also exhibit the potential stability of ~1.8 V. Efforts are on to measure

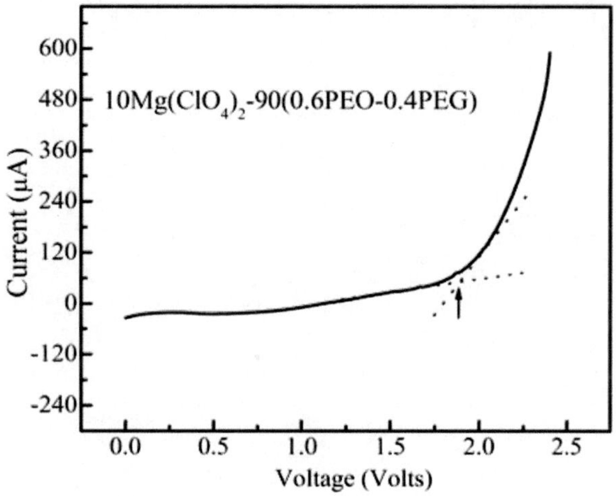

Fig. 8 LSV for the 10 $Mg(ClO_4)_2$-90(0.6PEO-0.4PEG)

the exact Mg^{2+} ion transport number and also to enhance Mg^{2+} ion transport by composition alterations.

References

1. Jaipal Reddy, M., Chu, P.P.: Effect of on PEO morphology and conductivity. Solid State Ionics **149**, 115–123 (2002)
2. Pandy, G.P., Agarwal, R.C., Hashmi, S.A.: Magnesium ion-conducting gel polymer electrolytes dispersed with fumed silica for rechargeable magnesium battery application. J. Solid State Electrochem. **15**, 2253–2264 (2011)
3. Mangalam, R., Thamilselvan, M., Selvasekarapandia, S., Jayakumar, S., Manjuladevi, R., Rangaswamy, S.: Development and study of solid polymer electrolyte based on polyvinyl alcohol: $Mg(ClO_4)_2$. Polymer Plast. Technol. Eng. **56**, 992–1002 (2017)
4. Mangalam, R., Thamilselvan, M., Selvasekarapandia, S., Jayakumar, S., Manjuladevi, R.: Polyvinyl pyrrolidone/$Mg(ClO_4)_2$ solid polymer electrolyte: structural and electrical studies. Ionics **23**, 2837–2843 (2017)
5. Anji, P.R., Ranveer, K., Rhee, H.-W.: Magnesium ion conducting solid polymer blend electrolyte based on biodegradable polymers and application in solid-state batteries. Ionics **21**, 125–132 (2015)
6. Anil kumar, K.M., Jinisha, B., Manoj, M., Jayalekshmi, S.: Poly(ethylene oxide) (PEO)—Poly(vinyl pyrrolidone) (PVP) blend polymer based solid electrolyte membranes for developing solid state magnesium ion cells. Eur. Polymer J. **89**, 249–262 (2017)
7. Anji, P.R., Ranveer, K., Valerio, C., Ramesh, N.: Conductivity, XRD, and FTIR Studies of new-ion-conducting solid polymer electrolytes: [PEG: $Mg(CH_3COO)_2$]. J. Korean Phys. Soc. **59**(1), 114–118 (2011)
8. Elashmawi, I.S., Gaabour, L.H.: Raman, morphology and electrical behavior of nanocomposites based on PEO/PVDF with multi-walled carbon nanotubes. Results Phys. **5**, 105–110 (2015)

Effect of Au Decoration on the Performance of Perovskite Solar Cells

Nilesh G. Saykar, Anil Arya, Indrani Banerjee, and S. K. Mahapatra

Abstract In the present work, TiO_2 as an electron transport layer (ETL) decorated with gold (Au) nanoparticles is demonstrated. It has been characterized by UV-Visible spectroscopy, FESEM, XRD prior to utilization as an ETL. The decoration of Au nanoparticles on the TiO_2 nanorods is confirmed through morphological analysis and depicts uniform decoration of Au nanoparticles while XRD analysis reveals the formation of rutile TiO_2 nanorods. The optimum concentration of Au nanoparticles for the device fabrication is achieved by examining device performance parameter such as V_{oc}, fill factor and power conversion efficiency. The plasmonic device shows enhanced performance as compared to bare TiO_2 devices. The bare TiO_2 exhibits 4.4% power conversion efficiency (PCE) while Au–TiO_2 have 6.68% highest PCE. The devices with Au–TiO_2 exhibits current density from 16.77 to 17.61 mA cm^{-2} with V_{oc} in the range of 0.82–0.86 V. Impedance spectroscopy analysis reveals that the decoration of Au nanoparticles reduces the charge transfer resistance. The obtained results reveal that the decoration of Au nanoparticles on the TiO_2 nanorods efficiently improves the performance of perovskite solar cells.

Keywords Perovskite solar cells · Au nanoparticles · Efficiency

1 Introduction

The current state-of-the-art of energy generation and storage technology is widely evolving due to the tremendous need for energy. There are several devices like solar cells and fuel cells being used to generate energy, while storage is done using Li-ion batteries and supercapacitors [1–3]. Among the various types of solar cells, perovskite solar cells (PSCs) have achieved a significant improvement in device technology. Still, there is a need for improvement in the performance of the PSC.

N. G. Saykar · A. Arya · S. K. Mahapatra (✉)
Department of Physical Sciences, Central University of Punjab, Bathinda 151001, India
e-mail: skmahapatra741973@gmail.com

I. Banerjee
School of Nano Sciences, Central University of Gujarat, Gandhinagar 382030, India

Y. Sharma et al. (eds.), *Recent Research Trends in Energy Storage Devices*,
https://doi.org/10.1007/978-981-15-6394-2_12

There are many approaches to improve the performance of perovskite solar cells by interface engineering, film formation, absorber design and defect passivation [4–6]. Interface engineering plays a vital role in these solar cells by controlling the charge carriers between different layers of the device. When plasmonic nanoparticles include in one of the layer either absorber or carrier transport layer resulting device performs better by use of a large amount of incident sunlight and engineered charge carrier dynamics. The enhancement is due to the reduced exciton binding energy in the device. Subsequently, perovskite solar cells also showed the plasmonic enhancement effects like improved charge transfer, light absorption and hot electron injection [7]. When localized surface plasmonic resonance (LSPR) happen, the cross-section of the nanocrystal and localized electric field will be highly improved which helps to confine the energy of incident light and induce photoelectric effects. So it is important to perform a fundamental and systematic study of plasmonic effect by controlling the metal geometry and metal-dielectric interface to boost up the power conversion efficiency (PCE) of the device.

Snaith et al. showed core-shell Au@SiO_2 nanoparticles in perovskite solar cells and achieved 11.4% PCE while in case of Ag@TiO_2 they achieved 16.3% PCE [8]. A highly efficient PSC by utilizing Au nanoparticles and MgO with 16.1% PCE is developed by Zhang et al. [9]. While the Au nanoparticles embedded TiO_2 nanofibers based PSC prepared by Mali et al. exhibits 14% PCE [10]. In the present work, we report an effective and facile method to prepare plasmonic-based PSC. This method is different from the previous report which contains bonding of Au nanoparticles to rutile TiO_2 nanorods without any functional group which promotes the fast transfer of charge carrier without intermediate steps.

2 Experimental

2.1 *Synthesis of TiO_2 Nanorods Using the Hydrothermal Method*

The TiO_2 nanorods are synthesized using the hydrothermal method reported previously [11]. In a typical process, double distilled water and concentrated hydrochloric acid (36.5–38%) 7 ml of each are mixed together. In this solution, 0.14 ml of titanium isopropoxide is added. The whole mixture stirred to dissolve the reactant completely. The FTO substrate is placed such a way that the FTO coated side is vertically down. After stirring, the mixture is placed in a Teflon-lined stainless steel autoclave of 25 ml volume and heated at 180 °C for 6 h. On the completion of the reaction, FTO substrate is washed 2–3 times with distilled water.

The decoration of TiO_2 nanorods with gold nanoparticles is carried out using $HAuCl_4$ as a precursor. In a typical process, 1 mg sodium borohydride is dissolved in 50 ml deionized water under continuous stirring at room temperature. A 100 μl Au precursor is added to the above solution from the stock solution (1.1 mM). After the preparation of Au nanoparticles, TiO_2 deposited FTO substrate is placed in it for various dipping time viz. 0.5 h, 1 h and 2 h namely Au–TiO_2 0.5 h, Au–TiO_2 1 h and Au–TiO_2 2 h, respectively.

2.2 Solar Cell Fabrication

Methyl ammonium lead iodide ($CH_3NH_3PbI_3$) is synthesized using the well-established method [12, 13]. In a typical process, Hydroiodic acid (HI) is added to methylamine in a 1:1.2 ratio with continuous stirring for 2 h in an ice bath. After completion of the reaction, the solvent is evaporated using a rotary evaporator. The precipitate is washed with diethyl ether. The $CH_3NH_3PbI_3$ solution is prepared by dissolving equimolar amounts of CH_3NH_3I and lead iodide (PbI_2) in dimethylformamide (DMF). This solution is spin-coated at 4000 rpm for the 30 s on TiO_2 and Au–TiO_2 samples subsequently heated for 10 min to form $CH_3NH_3PbI_3$. The hole transport layer (HTL) is prepared by the procedure [11]. The 180 mg 2,2′,7,7′- tetrakis-(N,N-di-p-methoxyphenyl-amine)-9,9′-spirobifluorene(spiro-MeOTAD) dissolved in 1 ml chlorobenzene with addition of 37.5 μl bis(trifiuoromethane)sulfonimide lithium (170 mg ml^{-1}) in acetonitrile and 17.5 μl 4-tert-butylpyridine. The $CH_3NH_3PbI_3$ deposited TiO_2 films are coated with HTL solution using the spin-coated at 3000 rpm for the 30 s. Then the substrates were transferred to a vacuum chamber and evacuated to a pressure of 2×10^{-6} mbar. For the Au-counter electrode, an 80 nm thick Au is deposited on the top of the HTL overlayer by thermal evaporation (growth rate ~0.5 Å/s) at a pressure of 2×10^{-6} mbar. The active areas of all devices are 0.2 cm^2.

3 Measurement

Absorption spectra are measured using UV-visible spectroscopy (Shimadzu uv-2450). The morphology of the samples is recorded from high-resolution FESEM (Carl Zeiss Merlin Compact). The structural analysis of samples is carried out through XRD spectra (PANanalytical). The current-voltage (J–V) characteristics of the prepared PSCs are measured using a Keithley source measuring unit (Model 2400) under AM 1.5 white light (100 mW cm^{-2}) illumination. Electrochemical impedance spectroscopy (EIS) measurements are performed using an impedance analyser (CHI 760). An EIS spectrum is obtained over a frequency range of 1 MHz to 100 Hz. The impedance parameters are obtained by fitting the impedance spectra using $Z_{simpwin}$ software.

4 Results and Discussion

The X-ray diffraction pattern of the TiO_2 nanorods is shown in Fig. 1. The experimental XRD pattern agrees with the JCPDS card no. 211276 (rutile TiO_2). Strong diffraction peaks at 27.77°, 38.84°, 55.66°, and 62.67° indicates that TiO_2 nanorods grow in the rutile phase corresponding to (110), (200), (220) and (002) plane. However, the peak at 28.1° and 35.23° correspond to (220) and (222) planes of Au nanoparticles having FCC structure, respectively, (JCPDS card no. 011174).

Figure 2a shows the UV-Vis spectra of gold nanoparticles. The presence of absorption peak at 520 nm reveals the formation of Au nanoparticles while the narrow peak indicates nanoparticles are monodisperse. Absorption spectra at different storage time reveal Au nanoparticles are stable up to 2 h. UV-Vis spectra of Au decorated TiO_2 nanorods shown in Fig. 2b. It shows that the decoration of Au enhances photon absorption.

Fig. 1 XRD spectra of **a** TiO_2, **b** Au–TiO_2 1 h

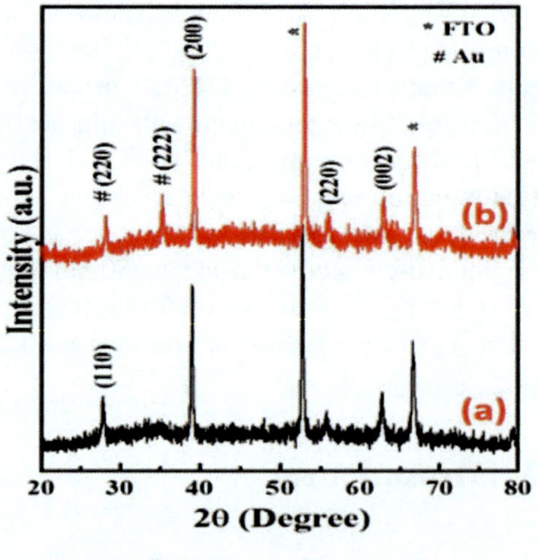

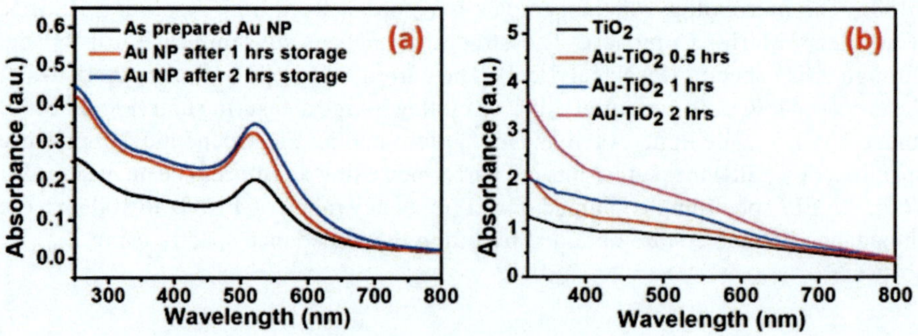

Fig. 2 **a** UV-Vis spectra of Au Nanoparticles, **b** Au–TiO_2 thin film

Fig. 3 FESEM images of **a** TiO_2, **b** Au–TiO_2 1 h, **c** Au–TiO_2 2 h, **d** perovskite deposited on TiO_2

The morphological analysis shows the TiO_2 grows in the rod-like morphology (Fig. 3a). From Fig. 3b, c it is seen that the TiO_2 nanorods are decorated with ~20 nm Au nanoparticles. Figure 5d shows the FESEM images of $CH_3NH_3PbI_3^-$ deposited on the TiO_2. From FESEM image it is seen that the $CH_3NH_3PbI_3$ uniformly crystallizes into larger grains. It has a grain size up to 1 µm which is important for recombination free charge transport.

The photocurrent density-voltage (J–V) characteristics of the fabricated devices with and without Au nanoparticles are investigated under AM 1.5G solar irradiance. Figure 4 shows the typical J–V plots for the devices with TiO_2 and Au−TiO_2 as an ETL. The TiO2 based devices exhibit 4.4% power conversion efficiency (PCE) while devices with Au–TiO_2 has a PCE from 5.66 to 6.68%. Au–TiO_2 used devices exhibit current density from 16.77 to 17.61 mA cm^{-2} with Voc in the range of 0.82–0.86 V and $FF = 35.98–39.93$.

The optimum concentration of Au for higher PCE is evaluated. From the JV measurement, it is seen that Au–TiO_2 1 h based devices have the highest efficiency. It is also found that Voc increases for Au–TiO_2 based perovskite solar cell. Further increase in Au concentration leads the decrement in PCE. The obtained results for solar cell parameters are recorded in Table 1.

The enhancement in Au-TiO_2 based devices is attributed to the plasmonic effect of Au nanoparticles. The schematic of device architecture is shown in Fig. 5a. The plasmonic coupling between Au nanoparticles promotes electron transfer into the

Fig. 4 Photocurrent density versus voltage measurement of solar cell

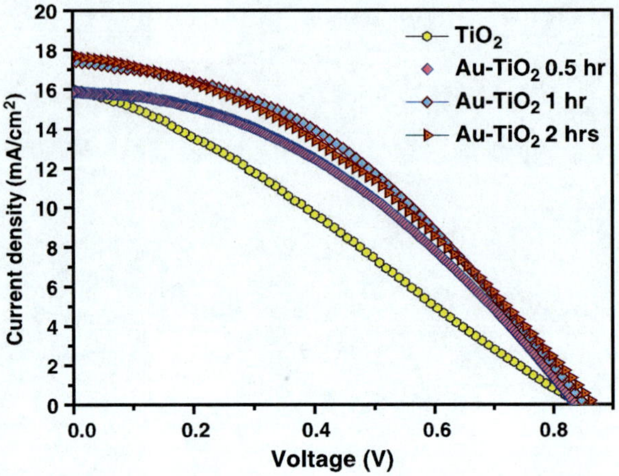

Table 1 Solar cell parameters

Sample	J_{sc} (mA/cm^2)	V_{oc} (V)	FF	η (%)
TiO_2	15.95	0.84	28.81	4.41
Au–TiO_2 0.5 h	16.77	0.82	35.98	5.66
Au–TiO_2 1 h	17.28	0.85	39.93	6.68
Au–TiO_2 2 h	17.61	0.86	37.13	6.4

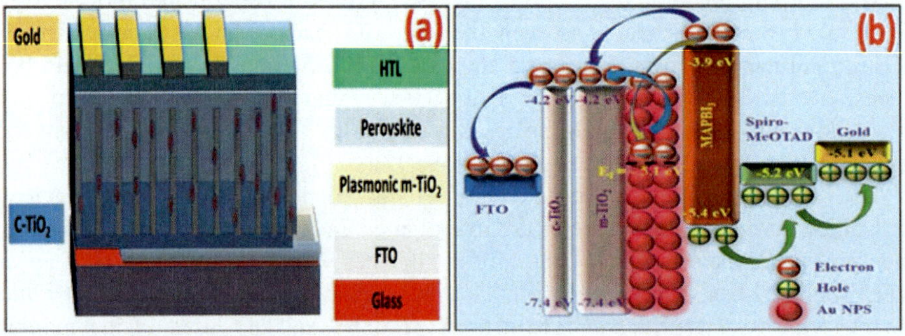

Fig. 5 **a** Schematic of device architecture, **b** schematic of energy levels of each layer of PSC

TiO_2. LSPR of Au nanoparticles helps in transfer of photoelectrons from SPR level to the conduction band (CB) of TiO_2 [10]. Figure 5b represents the typical energy band diagram of Au–TiO_2 based device. From this diagram, it is inferred that in Au based devices photon absorbed by the Au nanoparticles transferred to the CB of TiO_2 which increases the photon harvesting efficiency of the device as compared to bare TiO_2 based devices.

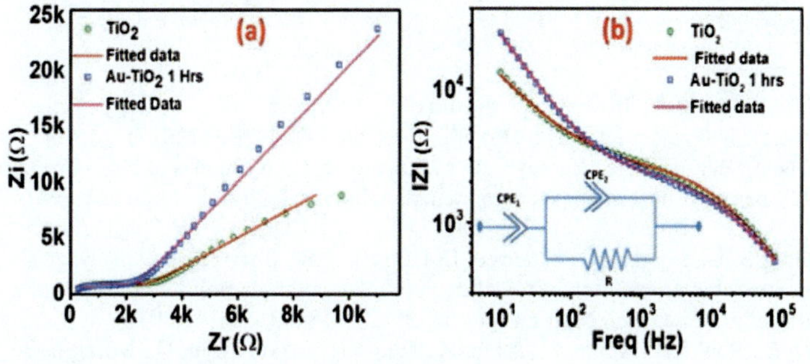

Fig. 6 EIS measurement of TiO_2 and Au–TiO_2 1 h based devices, **a** Nyquist plot, **b** variation of impedance with frequency of PSC, the inset shows the equivalent circuit

Table 2 Impedance fitting parameters

Sample	CPE_1 (S $sec^{0.5}/cm^2$)	R (Ω cm^2)	CPE_2 (S $sec^{0.5}/cm^2$)
TiO_2	7.56×10^{-6}	2435	1.29×10^{-7}
Au–TiO_2 1 h	1.67×10^{-6}	2220	1.71×10^{-7}

The electrochemical analysis of the fabricated solar cell is carried out and shown in Fig. 6. It shows the Nyquist plot of solar cells with and without Au nanoparticles under dark conditions at 0.3 V bias. It is seen that both devices exhibit semi-circle at a higher frequency corresponding to transfer resistance while another semi-circle at a lower frequency corresponding to recombination resistance. Incorporation of Au nanoparticles on to the ETL reduces the charge transport resistance which improves charge separation consequently the efficiency of PSC. The fitting parameters of devices are listed in Table 2.

5 Conclusion

Au decorated TiO_2 nanorods as an ETL for PSC has been successfully demonstrated. The optimum concentration of Au nanoparticles for higher PCE is evaluated. Au–TiO_2 1 h based PSC has the highest efficiency. The plasmonic coupling between Au nanoparticles promotes electron transfer into the TiO_2 which reduces transport resistance. It is also found that V_{oc} increases for Au–TiO_2 based perovskite solar cell which suggests the Au decorated TiO_2 nanorods can be used as an alternative ETL for plasmonic perovskite solar cells.

References

1. Arya, A., Sharma, A.L.: Insights into the use of polyethylene oxide in energy storage/conversion devices: a critical review. J. Phys. D Appl. Phys. **50**(44), 443002 (2017)
2. Saykar, N.G., Pilania, R.K., Banerjee, I., Mahapatra, S.K.: Synthesis of NiO–Co_3O_4 nanosheet and its temperature-dependent supercapacitive behavior. J. Phys. D Appl. Phys. **51**(47), 475501 (2018)
3. Mahapatra, S.K., Saykar, N., Banerjee, I., Hobson, P.R., Sharma, A.K., Ray, A.K.: Synthesis and dielectric characterisation of triiodide perovskite methylammonium lead iodide for energy applications. J. Mater. Sci. Mater. Electron. **29**(21), 18693–18698 (2018)
4. Saparov, B., Sun, J.P., Meng, W., Xiao, Z., Duan, H.S., Gunawan, O., Mitzi, D.B.: Thin-film deposition and characterization of a Sn-deficient perovskite derivative Cs2SnI6. Chem. Mater. **28**(7), 2315–2322 (2016)
5. Chen, Q., Zhou, H., Song, T.B., Luo, S., Hong, Z., Duan, H.S., Yang, Y.: Controllable self-induced passivation of hybrid lead iodide perovskites toward high performance solar cells. Nano Lett. **14**(7), 4158–4163 (2014)
6. Zhou, H., Chen, Q., Li, G., Luo, S., Song, T.B., Duan, H.S., Yang, Y.: Interface engineering of highly efficient perovskite solar cells. Science **345**(6196), 542–546 (2014)
7. Fan, R., Wang, L., Chen, Y., Zheng, G., Li, L., Li, Z., Zhou, H.: Tailored Au@ TiO_2 nanostructures for the plasmonic effect in planar perovskite solar cells. J. Mater. Chem. A **5**(24), 12034–12042 (2017)
8. Zhang, W., Saliba, M., Stranks, S.D., Sun, Y., Shi, X., Wiesner, U., Snaith, H.J.: Enhancement of perovskite-based solar cells employing core–shell metal nanoparticles. Nano Lett. **13**(9), 4505–4510 (2013)
9. Zhang, C., Luo, Q., Shi, J., Yue, L., Wang, Z., Chen, X., Huang, S.: Efficient perovskite solar cells by combination use of Au nanoparticles and insulating metal oxide. Nanoscale **9**(8), 2852–2864 (2017)
10. Mali, S.S., Shim, C.S., Kim, H., Patil, P.S., Hong, C.K.: In situ processed gold nanoparticle-embedded TiO 2 nanofibers enabling plasmonic perovskite solar cells to exceed 14% conversion efficiency. Nanoscale **8**(5), 2664–2677 (2016)
11. Mali, S.S., Shim, C.S., Park, H.K., Heo, J., Patil, P.S., Hong, C.K.: Ultrathin atomic layer deposited TiO_2 for surface passivation of hydrothermally grown 1D TiO_2 nanorod arrays for efficient solid-state perovskite solar cells. Chem. Mater. **27**(5), 1541–1551 (2015)
12. Burschka, J., Pellet, N., Moon, S.J., Humphry-Baker, R., Gao, P., Nazeeruddin, M.K., Grätzel, M.: Sequential deposition as a route to high-performance perovskite-sensitized solar cells. Nature **499**(7458), 316 (2013)
13. Jeon, N.J., Noh, J.H., Kim, Y.C., Yang, W.S., Ryu, S., Seok, S.I.: Solvent engineering for high-performance inorganic–organic hybrid perovskite solar cells. Nat. Mater. **13**(9), 897 (2014)

In-situ Fabrication of Alternately Stacked MoS_2/Au Multilayered Thin-Film Electrodes for Electrochemical Energy Storage Application

Pranjala Tiwari and Ramesh Chandra

Abstract Molybdenum disulfide (MoS_2) is considered as one of the most researched transition metal dichalcogenides in several energy harvesting and storage applications. Particularly, the layered structure of MoS_2 allows the intercalation and de-intercalation of ions through it, giving rise to its pseudocapacitive behavior which makes it a promising electrode material for electrochemical supercapacitors. In this work, we have synthesized alternately stacked MoS_2–Au–MoS_2 multilayered thin-film electrodes on stainless steel current collectors (1×1 cm^2) using DC magnetron sputtering technique. Electrochemical performance was evaluated by cyclic voltammetry, charge–discharge characteristics, and electrochemical impedance spectroscopy in neutral 1 M Na_2SO_4 aqueous electrolyte using CHI workstation in 3 electrode configuration. The areal capacitance of bare MoS_2 and MoS_2–Au–MoS_2 multilayered electrodes was found to be 198 and 246 mF/cm^2 respectively, as calculated from the voltammetry graphs at the scan rate of 5 mV/s at similar mass loadings. Moreover, the multilayered electrodes displayed excellent cyclic retention capacities of 97.6% in comparison to single-layer MoS_2 whose capacity faded to 71% after 1500 cycles. This excellent structural stability response may be credited to the presence of sandwiched thin Au layer which provides an opportunity of epitaxial growth of thin films, subsequently minimizing uncompensated thermal and lattice stress.

Keywords Thin films · Sputtering · MoS_2 · Supercapacitor

1 Introduction

Ultrathin layered transition metal dichalcogenides have evolved as a generic class of materials which has gained significant consideration in diverse applications because of their intriguing chemical and electrical properties varying from metallic to semiconductor behavior [1]. Especially MoS_2, due to its known high intrinsic ionic

P. Tiwari · R. Chandra (✉)
Nanoscience Laboratory, Institute Instrumentation Centre, Indian Institute of Technology-Roorkee, Roorkee 247667, India
e-mail: ramesh.chandra@ic.iitr.ac.in

Y. Sharma et al. (eds.), *Recent Research Trends in Energy Storage Devices*,
https://doi.org/10.1007/978-981-15-6394-2_13

conductivity and large theoretical capacity comparable to that of graphite, has been a part of multiple research groups focusing on energy storage applications [2]. The layered sheet-like morphology of MoS_2 along with the wide interplanar spacing of 0.615 nm provides a considerable large surface area, contributing to high double-layer capacitance. Molybdenum also exhibits variable oxidation states in the range of +2 to +6 similar to that of oxides which allows pseudocapacitive charge transfer on the Mo metal centre at lower sweep values, giving rise to its reversible intercalation characteristics [3]. In addition, the layered structure of MoS_2 with weak Van der waal forces between the basal planes allows the intercalation of electrolytic ions in the adjacent layers [4]. Accordingly, the charge storage phenomenon in MoS_2 can potentially due to both EDLC and pseudocapacitive processes [5]. However, poor electronic conductivity and limited cyclic stabilities of MoS_2, is still one of the crucial problems which limit its maximum specific capacitance value. In order to boost the electrochemical properties of MoS_2, quite a few researchers have actively adopted various surface modifications and doping techniques employing transition metal compounds as well as conductive polymers [6–9].

Another limitation is the weak structural stability of multilayered 2H–MoS_2, which is primarily due to the lattice mismatch between the film and the current collector. Moreover, thermally induced stress is also generated at the interface of the film and the substrate during the cooling time after the growth due to dissimilarities between the thermal expansion coefficients of both. This stress and misfits contribute significantly to the interfacial adhesion and structural stability of films having a thickness in the submicron scale range if the lattice mismatch is greater than ~10%. Beyond this 10% range, this mismatch in material properties cannot be housed by the elastic lattice strain alone and it may give rise to various forms of dislocations at the interface in order to release the interfacial strain [10, 11]. These dislocations may further contribute to degrading the structural integrity during electrochemical analysis. One approach to tackle this problem is hierarchically stacking a conductive metallic layer such as Au, between two adjacent layers of MoS_2 to form a multilayered electrode. Several studies have claimed the similarities in the lattice parameters of 2H–MoS_2 and Au, resulting in strong epitaxial growth of MoS_2 on Au and vice versa [12, 13]. Moreover, the idea behind the stacking of the thin metallic layer lies in the fact that this layer can greatly enhance the intrinsic conductivity of the electrode which accelerates the electron conduction as well as the proton diffusivity in the electrode [14]. Moreover, Au would also contribute to the pseudocapacitive behavior of MoS_2 and enhance charging–discharging rates.

The sputtering technique can be used to optimize thin films in a wide nanometer regime more rapidly than exfoliation or solution phase methods, providing excellent control over the film's thickness, stronger adhesion and with the possibility of scale-up and mass production. In this work, we present a simple scalable technique to fabricate high-performance MoS_2/Au multilayered electrode using a direct current (DC) magnetron technique using commercially available MoS_2 and Au targets (99.9% pure). The sputtering technique allows forming a unique multilayered thin-film structure with inevitable interfacial contact between MoS_2 and Au which enables much faster ionic as well as electronic transport throughout the electrode matrix along with maximizing the available surface area sites by exposing the substrate electrode completely for capacitive charge storage.

2 Experimental Details

2.1 Synthesis

Multilayered MoS_2–Au electrode fabrication was performed using DC magnetron sputtering using a commercially available MoS_2 target and Au target (99.9% purity) directly onto stainless steel (SS) substrate. Prior deposition, SS (~2 Ω) was cleaned using ethanol, and then cut into 3 × 1 cm pieces for electrode preparation. 2 × 1 cm^2 area was masked and the electrodes were directly fabricated onto 1 × 1 cm^2 SS by sputtering. Initially, chamber was evacuated to a vacuum of 3×10^{-6} Torr and the substrate temperature was maintained at 300 °C. The deposition of base MoS_2 layer on SS was taken by adjusting the DC power to 60 W at base pressure of 5 mTorr for 3 min. Next, an ultrathin layer of Au was deposited onto the sputtered MoS_2 layer at 50 W and 20 Torr base pressure for 60 s. And finally, the top MoS_2 layer was deposited again by following the same parameters for the bottom layer. The electrode was weighed before and after the deposition of multilayered MoS_2–Au–MoS_2 thin films.

2.2 Characterizations

The X-ray investigation of prepared multilayered electrodes was performed using Bruker D8 advanced diffractometer. In order to examine the morphology as well as quantitative elemental analysis, FE-SEM (Carl Zeiss, Ultra plus) and EDAX (Oxford Instruments) were used, respectively. The avg. roughness of the film was explored using atomic force microscopy (NTMDT NTEGRA) functioned in semi-contact mode. Chemical state of the samples was characterized and confirmed using PHI Versa Probe III X-ray photoelectron spectrophotometer (XPS). The charge storage properties were studied using CHI660D Potentiostat with a voltage window of 0 to −0.8 volts in 1 M Na_2SO_4 electrolyte. Multilayered MoS_2–Au thin films were made as a working electrode, Pt as counter electrode, and Ag/$AgCl_2$ as reference electrode to study the three-electrode characteristics. Charge–discharge cycles were also measured for the respective samples in order to analyze the columbic efficiency and IR losses.

3 Results and Discussion

The x-ray diffraction (XRD) results of the Au/MoS_2 electrodes fabricated at 300 °C is shown in Fig. 1, confirming the growth of 2H–MoS_2 structure matched to JCPDS card no. 00-002-1133.

The major peaks located at 34.05, 43.05, and 50.54° correspond to hexagonal MoS_2 while the peak located at 43.86° corresponds to (200) plane of Au (JCPDS no 78-0430). Morphological results shown in Fig. 2a indicates dense forest of well-distributed MoS_2 nanoflakes uniformly all over the substrate. Images were taken at higher magnification as shown in Fig. 2b, c indicates that the size of flakes is 80–150 nm while the width is 10–30 nm. The cross-section of film was studied and is shown as an inset of image 2c, indicating that the thickness of the deposited electrode is about 110 nm. The elemental mapping results taken by EDAX are shown in Fig. 2d–f, indicating Mo, S and Au in the active material. Figure 3a–c displays the 2D AFM images of fabricated electrodes at the scanned area of about 10 × 10, 5 × 5 and 2 × 2 μm^2. The root mean square roughness of multilayered film was found to be ~7.62 nm as shown in the 3D image in Fig. 3d. The XPS survey scan is depicted in Fig. 4a and the $C - 1$ s spectra (Fig. 4b) is used for the calibration of other elements located at 248.8 eV. The high-resolution elemental spectra are used to determine the respective oxidation states and are shown in Fig. 4b–e for Mo 3d, Au-4f, $C - 1$ s, and S 2p, respectively. The Mo 3d spectrum can be fitted with a doublet with the Mo $3d_{5/2}$ positioned at 228.5 eV and Mo $3d_{3/2}$ positioned at 231.71 eV (orbital splitting distance of 3.2 eV). The third peak at a lower binding energy of 225.6 eV is indicative of S 2 s orbital. The S 2p core level displays two doublets peaks positioned at 162.15 and 162.44 eV, assigned as S $2p_{3/2}$ and S $2p_{1/2}$ (Fig. 4c). The high-resolution spectra of Au-4f are shown in Fig. 4e, de-convoluted into 4 $f_{7/2}$ and 4 $f_{5/2}$ peaks.

Further, we have studied the electrochemical properties of the as-prepared Au/MoS_2 and single-layer MoS_2 electrode material for supercapacitors. The contribution from the ultrathin Au layer is almost negligible due to its small sputter time of

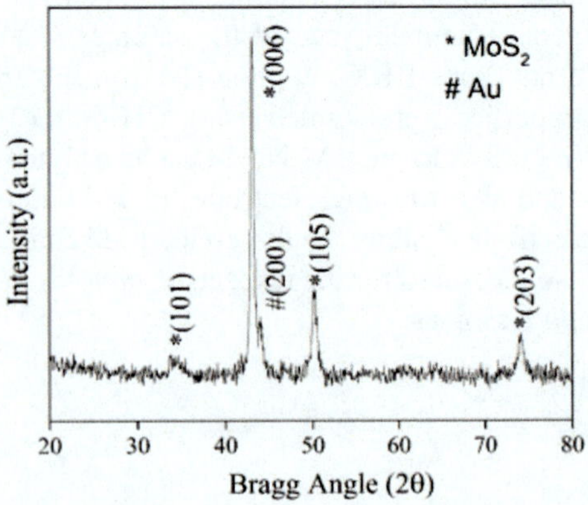

Fig. 1 XRD results depicting MoS_2/Au multilayered electrodes

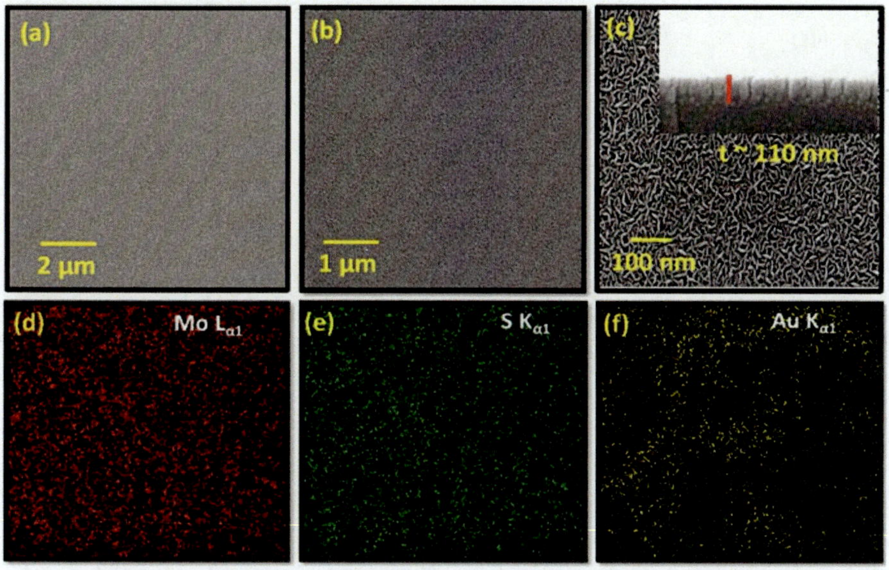

Fig. 2 **a–c** SEM images depicting morphology of multilayered electrode at various magnification scales, **d–f** elemental mapping showing the presence of Mo, S and Au in respective proportions

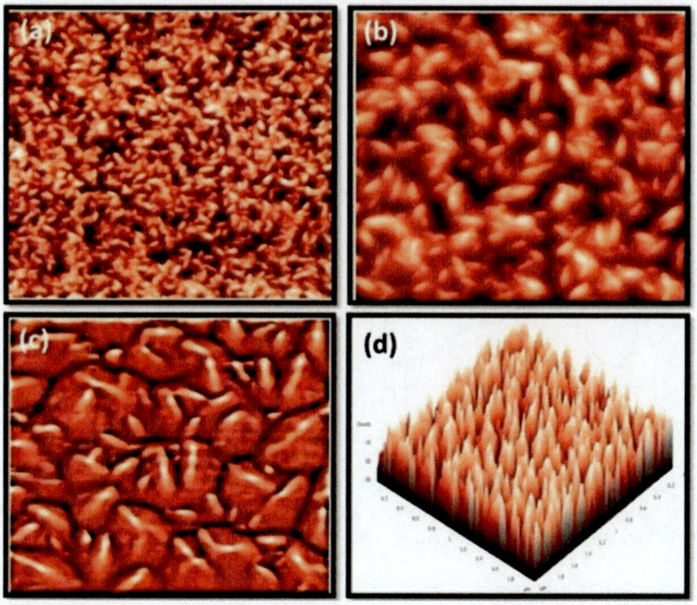

Fig. 3 2D images of the electrode at scanned in area 10 × 10 cm^2 (**a**), 2 × 2 cm^2 (**b**), 1 × 1 cm^2 (**c**) and 3D AFM image depicting the average RMS roughness of the film (**d**)

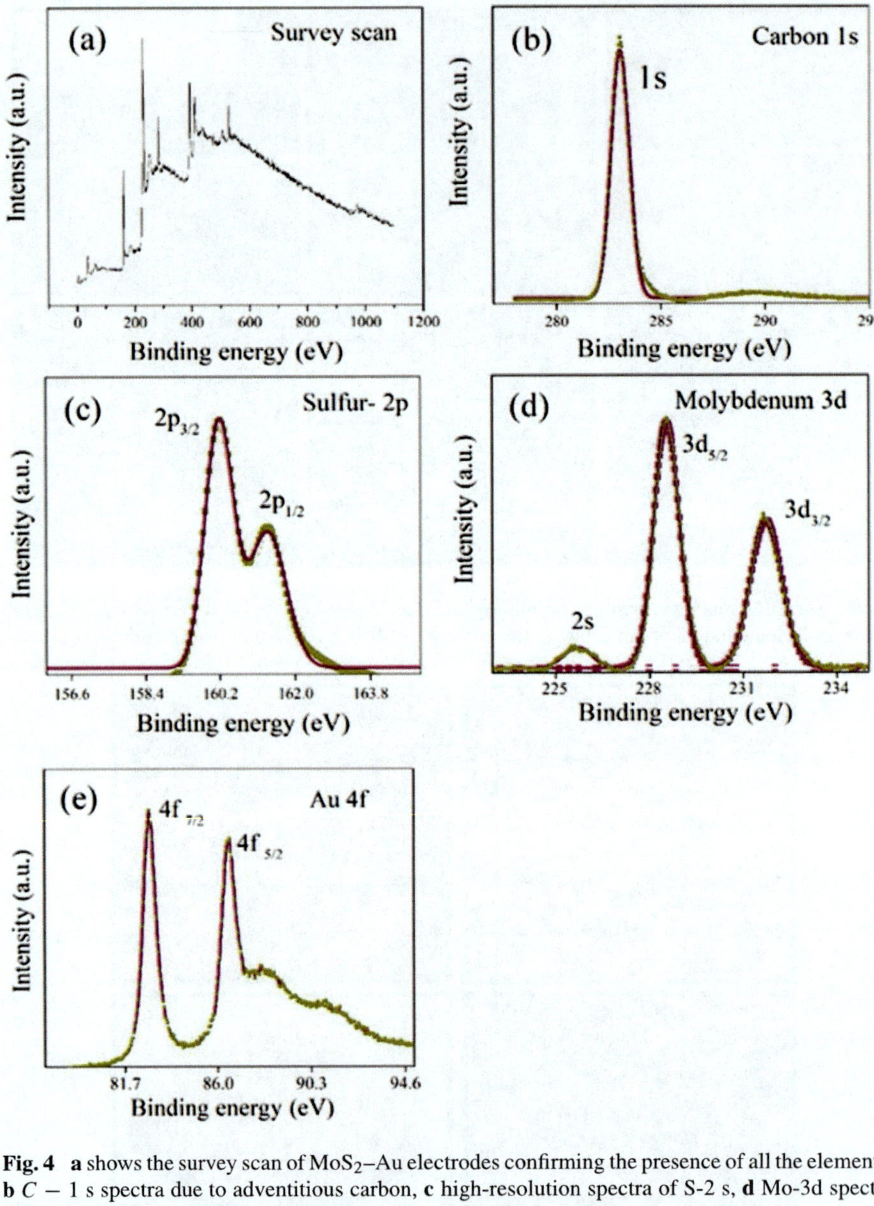

Fig. 4 **a** shows the survey scan of MoS_2–Au electrodes confirming the presence of all the elements, **b** $C - 1$ s spectra due to adventitious carbon, **c** high-resolution spectra of S-2 s, **d** Mo-3d spectra and **e** shows high-resolution spectra of Au-4f core level

30 s, which can be seen in Fig. 5b. Moreover, since it is not a top layer, its contribution to the total capacitance could be assumed to be negligible. Also, it is worth noticing that the mass loading of both the electrodes is almost similar, yet the performance of the multilayered electrode is much better. Figure 5b shows the cyclic voltammetry curves for electrode at potentials in the range of −0.8 V to 0 V versus the Ag/AgCl, serving as excellent negative electrode material at various scan rates. The charge storage mechanism as depicted from the shape indicates dominantly towards characteristic EDLC behavior. However, the deviance from the ideal rectangular curve signifies the presence of some amount of pseudocapacitance also, which may be due to reversible redox reactions at the surface or due to the intercalation of Na^+

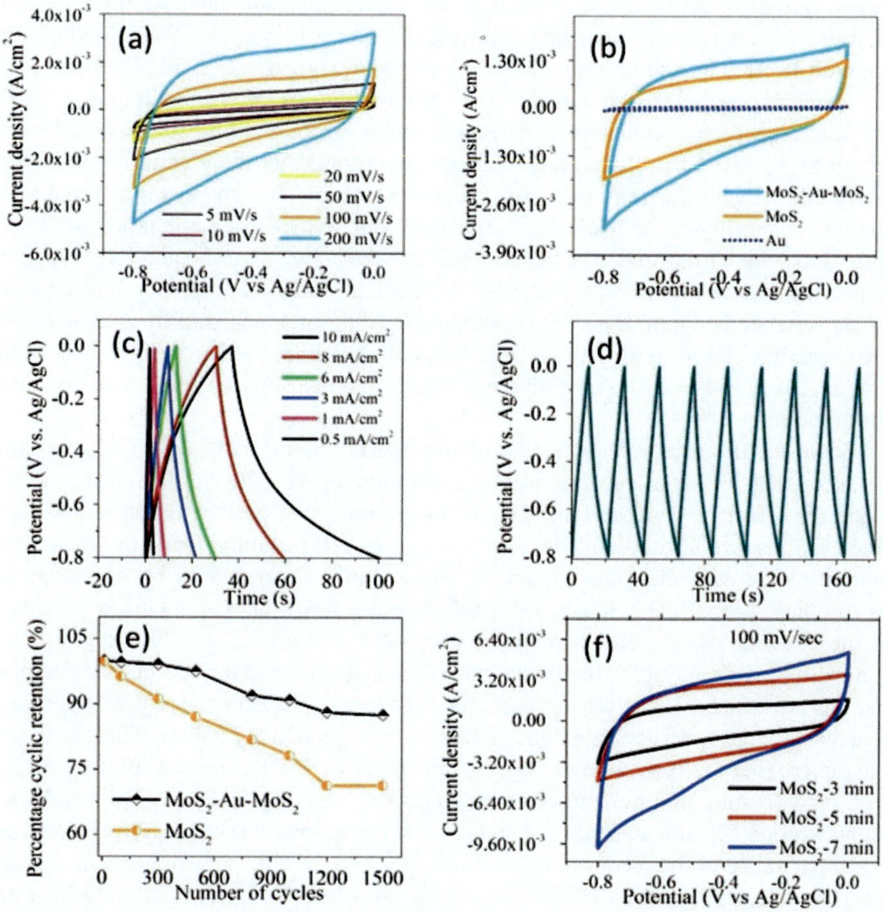

Fig. 5 Three electrode measurements of multilayered electrodes: **a** CV of MoS_2–Au at different scan rates, **b** shows comparative CV curve obtained for bare MoS_2 and MoS_2/Au electrodes, **c** GCD curves measured at different current densities, **d** shows charging–discharging curves of last cycles, **e** effect of cycling on percentage capacitance and **f** effect of mass loading on the specific capacitance

ions into MoS_2/Au matrix. The superior performance may be attributed to the fact that the sandwiched Au thin films enhance the intrinsic conductivity of the electrode while providing new interstices for the intercalation of electrolytic ions. GCD profiles of the same were acquired at various current densities (0.5–10 mA/cm^2) as shown in Fig. 5c. The presence of non-linearity in triangular graphs again advocates the contribution of both faradic and non-faradic mechanisms. The morphology depicts that the thin film grows exhibiting a porous structure. This morphology provides a higher number of sites for ionic diffusion as well as electrode-electrolyte interaction for electrostatic adsorption for double-layer charge storage. The areal capacitance of bare MoS_2 and MoS_2–Au–MoS_2 multilayered electrodes was found to be 198 and 246 mF/cm^2 respectively, as calculated from the voltammetry graphs at the sweep rate of 5 mV/s. In order to study the cyclic retention capacity after several charging–discharging cycles, we performed CV for 1500 cycles at 50 mV/s for both the multilayered and single-layer MoS_2 electrode as shown in Fig. 5d. We found that the single-layered electrode has cyclic retention of 70.8% after completion of 1500 CV cycles, which demonstrates its major limitation hampering the performance. Whereas, in case of multilayered electrode, we found that the electrode maintains an excellent cyclic capacity of 97.6% after 1500 cycles. This major improvement in the cyclic stability of the MoS_2/Au electrode is due to the strong epitaxial growth of thin films which minimizes the peeling effect of the active material and improves the structural integrity of films during electrochemical performances. The lower value of capacitance is due to the very small mass loading of MoS_2 thin films. The SEM cross-section images of the films suggest that the thickness of MoS_2 films is around 250 nm (inset of Fig. 2c), in which it was difficult to measure the small mass loading differences.

However, the principle objective of this work was to study the effects of Au, as an interfacial layer, on the structural stability of MoS2 electrodes. Moreover, since the effect of structural stability is more prominent on thin films, rather than thicker films, we limited the film thickness up to 110 nm only. Once the structural stability improves, the mass loading of upper MoS_2 layers could be increased so as to obtain higher capacitance values. Figure 5e shows the GCD curves obtained at the last nine cycles after completion of 1500 full charging–discharging sweeps, indicating good columbic efficiencies and rate capability. As far as commercial applications are concerned, relatively high mass loading of materials is required in order to achieve large specific capacitance as well as energy density values. Therefore, we have increased the thickness of upper MoS_2 layer by increasing the sputter time to 5, 8, and 10 min, and performed their cyclic voltammetry analysis. The results as shown in Fig. 5f, indicates that as the thickness (mass loading) of the films increases, the capacitance of the electrodes also increases. The specific capacitance was calculated as 282, 259, and 245 F/g for samples prepared at 3 min, 5 min, and 8 min respectively. We found that with the increase in mass loading the areal capacitance increases while the specific capacitance decreases. Though Au is an expensive material, the deposition of Au was taken only for 60 s to deposit an ultrathin layer, so as to improve the mechanical strength of the top layer. The reason behind the selection of

Au as an interfacial layer lies in the similarity in the lattice parameters and thermal conductivity of MoS_2 and Au.

However, Au can also be replaced by other materials that possess similar lattice parameters as that of MoS_2. Besides Au, other materials such as hexagonal boron nitride, Cu (111), CuCl (200) are known to have closed lattice matching with 2H–MoS_2 structure [15–17]. Thus, they may be explored as well, to improve the structural strength and subsequently, the cyclic capacity of MoS_2 electrodes.

4 Conclusions

This enhanced performance of multilayered thin-film electrodes may be attributed to the presence of thin Au layer between two adjacent MoS_2 layers which effectively increases the intrinsic electronic conductivity of MoS_2, while maintaining the same mass loading. This is in accordance with the fact that a highly conductive electrode can achieve better polarization and faster charge transport mechanism even at lower electrode mass loadings. Moreover, the electrode stability was assessed for 1500 charge–discharge cycles indicating 97.6% capacitive retention for multilayered electrodes which was almost 26% higher than that of bare MoS_2 electrodes. This indicates that apart from enhancing the intrinsic conductivity, the Au layer also improves the structural stability of MoS_2 thin films and thus slowing down the electrochemical dissolution of the electrode in the electrolyte.

Acknowledgements The author Pranjala Tiwari would like to acknowledge financial support from DST INSPIRE fellowship with grant number 2017/IF170545.

References

1. Guo, Y., Sun, D., Ouyang, B., Raja, A., Song, J., Heinz, T.F., Brus Louis, E.: Probing the dynamics of the metallic-to-semiconducting structural phase transformation in MoS_2 crystals. Nano Lett. **15**, 5081–5088 (2015)
2. Huang, K.J., Wang, L., Liu, Y.J., Liu, Y.M., Wang, H.B., Gan, T., Wang, L.L.: Layered MoS_2 graphene composites for supercapacitor applications with enhanced capacitive performance. Int. J. Hydrogen Energy **38**, 14027–14034 (2013)
3. Rasamani, K.D., Alimohammadi, F., Sun, Y.: Interlayer-expanded MoS_2. Mater. Today **20**, 83–91 (2017)
4. Zhang, H., Loh, K.P., Sow, C.H., Gu, H., Su, X., Huang, C., Chen, Z.K.: Surface modification studies of edge-oriented molybdenum sulfide nanosheets. Langmuir **20**(16), 6914–6920 (2004)
5. Soon, J.M., Ping, K.: Electrochemical double-layer capacitance of MoS_2 nanowall films. Electrochem. Solid-State Lett. **10**(11), 250–254 (2007)
6. Ma, L., Zhou, X., Xu, L., Xu, X., Zhang, L.: Microwave-assisted hydrothermal preparation of SnO_2/MOS_2 composites and their electrochemical performance. NANO **11**(2), 16–18 (2016)
7. Luo, W., Zhang, G., Cui, Y., Sun, Y., Qin, Q.: One-step extended strategy for the ionic liquid-assisted synthesis of Ni3S_4–MoS_2 heterojunction electrodes for supercapacitors. J Mater Chem A Mater Energy Sustain **5**, 1–8 (2017)

8. Geng, X., Geng, X., Yelong, Z., Han, Y., Li, J., Yang, L., Benamara, M., Chen, L., Zhu, H.: Two-dimensional water-coupled metallic MoS_2 with nanochannels for ultrafast supercapacitors. Nano Lett. **17**(3), 1825–1832 (2017)
9. Huang, K., Wang, L., Liu, Y., Wang, H., Liu, Y., Wang, L.: Synthesis of polyaniline/2-dimensional graphene analog MoS_2 composites for high-performance supercapacitor. Electrochim. Acta **109**, 587–594 (2013)
10. Bhattacharyya A, Maurice D (2018) On the evolution of stresses due to lattice misfit at a Ni-superalloy and YSZ interface. Surf. Interf.
11. Men, H., Fan, Z.: Effects of lattice mismatch on interfacial structures of liquid and solidified Al in contact with hetero-phase substrates: MD simulations. IOP Conf. Ser. Mater. Sci. Eng. **27**, 012007 (2012)
12. Grønborg, S.S., Ulstrup, S., Bianchi, M., Dendzik, M., Sanders, C.E., Lauritsen, J.V., Hofmann, P., Miwa, J.A.: Synthesis of epitaxial single-layer MoS_2 on Au(111). Langmuir **31**(35), 9700–9706 (2015)
13. Zhou, Y., Kiriya, D., Haller, E.E., Iii, J.W.A., Javey, A., Chrzan, D.C.: Compliant substrate epitaxy: Au on MoS2. Phys. Rev. B **93**, 054106 (2016)
14. Chen, T., et al.: All-solid-state high performance asymmetric supercapacitors based on novel MnS nanocrystal and activated carbon materials. Sci. Rep. **6**, 23289 (2016)
15. Kong, B.D., Zeng, C., Gaskill, D.K., Wang, K.L., Kim, K.W.: Two dimensional crystal tunneling devices for THz operation. Appl. Phys. Lett. **101**, 263112 (2012)
16. Wang, S., et al.: Epitaxial templating of two-dimensional metal chloride nanocrystals on monolayer molybdenum disulphide. ACS Nano **116**, 6404–6415 (2017)
17. Zhang, F., et al.: Full orientation control of epitaxial MoS_2 on hBN assisted by substrate defects. Phys. Rev. B **99**, 155430 (2019)

Conductivity and Dielectric Spectroscopy of Na^+ Ion Conducting Blended Solid Polymer Nanocomposites

Pritam, Anil Arya, and A. L. Sharma

Abstract A new blend solid polymer electrolyte based on polyethylene oxide (PEO)-poly (vinyl pyrrolidone) (PVP) with sodium nitrate ($NaNO_3$) has been synthesized via the standard solution cast technique. The temperature-dependent dielectric properties of the prepared polymeric films have been investigated by dielectric spectroscopy. With the increase of temperature the dielectric constant and dc conductivity increases. The increase in dielectric strength with temperature for polymer salt matrix (4.27×10^7 at 40 °C to 1.59×10^8 at 100 °C) evidences the enhancement in the number of charge carriers. The frequency associated with the maximum polarization shifts toward the high frequency and infers the effective role of salt and temperature. Then the relaxation time is obtained that decreases with the increase of temperature (57 s; at 40 °C to 25 s; at 100 °C) and are associated with the faster segmental motion of the polymer chain. A briefly discussed model has been proposed to justify the obtained experimental results.

Keywords Polymer electrolytes · Dielectric constant · Conductivity

1 Introduction

Due to the reduction of traditional energy resources such as coal, fossil fuel, etc., the development of a renewable source of energy became the foremost necessity across the world. The success like development on the solar cell, wind energy, fuel cell, etc. are going on but the parallel concern is there that how to store it. Therefore supercapacitor and secondary/rechargeable batteries seem to be more convincing and justified [1, 2]. In the availability approach of a different class of secondary batteries, Li and Na ion batteries are growing very fast. As the least abundance of Li on the earth surface, the major thrust is driving towards the development of Na-ion batteries [3]. In any of secondary batteries, the electrolyte is a key component, but presently used conventional electrolytes are liquid in nature that possess drawbacks such as

Pritam · A. Arya · A. L. Sharma (✉)
Department of Physics, Central University of Punjab, Bathinda 151001, Punjab, India
e-mail: alsharma@cup.edu.in

Y. Sharma et al. (eds.), *Recent Research Trends in Energy Storage Devices*,
https://doi.org/10.1007/978-981-15-6394-2_14

safety issues, a narrow temperature range of operation. In order to overcome the abovementioned issue, solid polymer electrolytes (SPEs) seem to be the most fascinating candidate for application due to high flexibility, improved stability properties and safety as compared to the conventional liquid electrolytes [4–7].

A number of the polymers have been tried as a host matrix for the development of such electrolytes. But Polyethylene oxide (PEO) is the most commonly used host polymer owing to the presence of an electron-rich ether group that provides coordinating sites to the cation for the migration and lower glass transition temperature. At the same time, its semi-crystalline nature restricts its use due to low electrical conductivity at room temperature. The most common approach to suppress the crystallinity is blending and is a feasible approach to improve the electrical and mechanical properties. After a critical review, polyvinyl pyrrolidone (PVP) has been chosen as a blend polymer due to high amorphous content and better compatibility with the chosen host polymer. It seems that after blending the amorphous nature of the host polymer would be improved [8, 9]. Also, with the increase of the temperature, the segmental motion may be improved attributed to the smooth and amorphous nature of the polymeric electrolyte. The ion dynamics in SPE is investigated by exploring the frequency of dependent dielectric permittivity and conductivity.

The present paper reports, the preparation of free-standing blend solid polymer electrolyte (BSPE) comprising of PEO and PVP, and $NaNO_3$ and ion dynamics with temperature. The effect of temperature on the complex conductivity, dielectric constant, dielectric loss, and relaxation time has been discussed in detail to understand the ion transport mechanism.

2 Materials and Methodology

2.1 Materials Used

The polyethylene oxide (PEO; MW: 6 × 10^5 g/mol), polyvinylpyrrolidone (PVP: MW 1.5 × 10^5 g/mol), are used as polymer, sodium nitrate ($NaNO_3$) as salt and methanol as solvent. All materials were purchased from Sigma Aldrich. Figure 1 shows the chemical structure of the materials used.

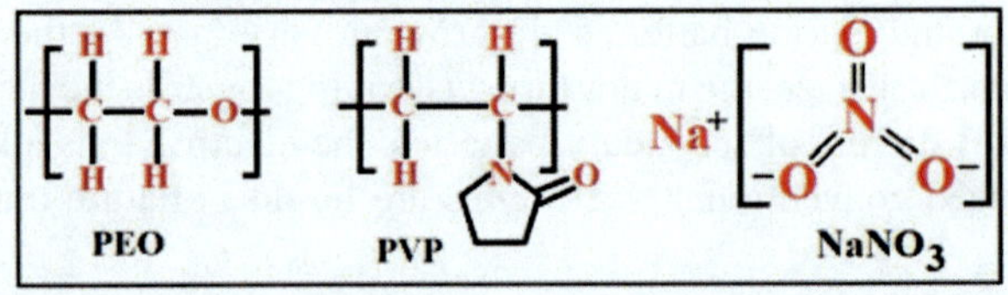

Fig. 1 Chemical structure of polyethylene oxide (PEO), polyvinylpyrrolidone (PVP) and sodium nitrate ($NaNO_3$)

2.2 Sample Preparation Methodology

For preparing blend solid polymer electrolytes, the standard solution cast technique has been adopted. Initially, the PEO-PVP (80:20) was added in methanol (20 ml) and kept on the stirring at room temperature for 4–5 h till a homogenous solution is obtained. Then the appropriate optimized stoichiometric ratio of salt (O/Na = 14) was added in the polymer solution and stirring is done for 10–12 h. After that, the final viscous solution is cast in petri-dish and kept at room temperature for evaporation of the solvent. Finally, a free-standing solid polymer electrolyte film is obtained. The average thickness of the prepared film is ~100 m.

2.3 Characterization Technique

The dielectric spectroscopy is an important technique for obtaining information about the complex permittivity, complex conductivity, and relaxation time of the polymeric system. The dielectric parameters have been obtained from the impedance spectroscopy. The impedance spectroscopy has been performed using the electrochemical analyzer (CHI 760; USA) in the frequency range of 1 Hz–1MHz and temperature range 40–100 °C (Temperature controller; Marine India). An ac signal of 20 mV is applied across the cell configuration SS|BSPE|SS. Where SS refers to stainless steel electrodes and BSPE is the blend solid polymer electrolyte. For dielectric analysis further, the impedance data transferred into the dielectric format. The dielectric properties of synthesized blend solid polymer electrolyte have been investigated in the term of complex conductivity and complex permittivity. And also complex permittivity and complex conductivity plots were fitted through established theoretical expression in the frequency window to get the proper insight of dielectric relaxation and ion dynamics in the blend solid polymer electrolyte [10, 11].

3 Results and Discussion

3.1 Complex Dielectric Permittivity Analysis

The complex dielectric permittivity for a solid polymer electrolyte film is expressed as $\varepsilon^* = \varepsilon' - j\varepsilon''$, where ε' is the real part and ε'' is imaginary part of dielectric permittivity. The real part of the complex dielectric permittivity (ε') is associated with stored energy and an imaginary part of dielectric permittivity (ε'') depicts the loss of energy within the medium [12–14]. Figure 2a, b shows the frequency-dependent variation of the real part of complex permittivity for the pure blend and polymer salt matrix. It is observed that the dielectric constant decreases with an increase of frequency, followed by steady-state at a higher frequency. The high value of

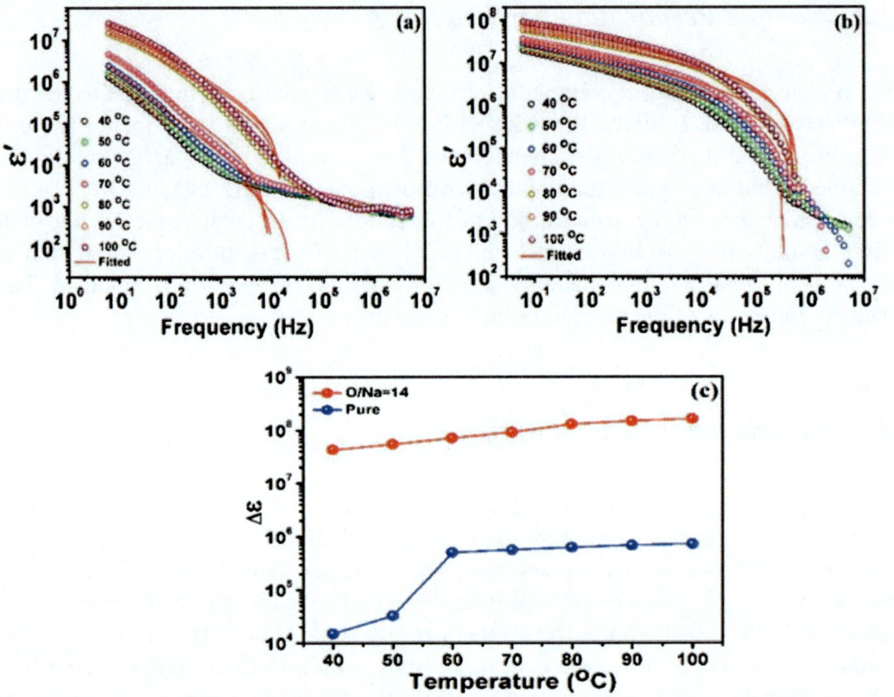

Fig. 2 The frequency and temperature-dependent real part of dielectric spectroscopy, **a** pure PEO/PVP (b) PEO/PVP + O/Na = 14. The solid red line is fitted plot and **c** temperature dependence of dielectric strength

the dielectric constant in the low-frequency window is attributed to the electrode polarization region and decrease at high frequency is associated with the dispersion region. Figure 2c shows the variation of the dielectric strength with temperature for the pure blend and polymer salt system. The addition of the salt increases the dielectric strength and agrees well in the broad temperature range (40–100 °C) [15–18]. The value of dielectric strength for the polymer salt matrix increases from 4.27 $\times 10^7$ (at 40 °C) to 1.59 $\times 10^8$ (at 100 °C).

The increase of temperature results in better salt dissociation that increases the polarization owing to the increased number of free charge carriers according to the relation: $\sigma_i = \sum_i q_i n_i \mu_i$ where σ_i is the conductivity, q_i is a charge, n_i is the number of charge carriers and μ_i is the mobility.

Figure 3a, b shows the frequency and temperature-dependent imaginary part of dielectric permittivity for solid polymer electrolyte film (pure blend and polymer salt). The dielectric loss decreases with the increase of frequency. The dielectric loss at low frequency is dominated by the electrode polarization and space charge polarization that generates heat inside the dielectric [19, 20]. In all sample, the dielectric loss increase with the increase of temperature, and there no maxima peak is observed.

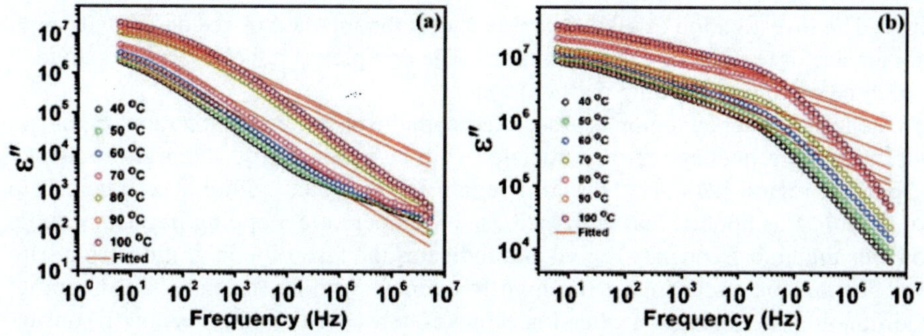

Fig. 3 The frequency and temperature-dependent imaginary part of complex dielectric permittivity, **a** pure PEO/PVP, **b** PEO/PVP + O/Na = 14. The solid red line is fitted plot

The increase in loss is due to the increased number of charge carriers. Both pure blend and polymer salt systems follow the same trend. The experimental results are fitted with the theoretical models proposed by Roy et al. [11]. It is important to note that the experimental results are in agreement with the theoretical equations. The small deviation at high frequency is due to the dominance of the relaxation processes.

3.2 Complex Conductivity Analysis

The complex conductivity is expressed as $\sigma^* = \sigma' - j\sigma''$, where σ' is the real part and σ'' is imaginary part of complex conductivity. The temperature-dependent real and imaginary part of complex conductivity is shown in Fig. 4 a, b for the pure blend and polymer salt film. Figure 4a, b shows the plot of the real part of conductivity against the frequency at different temperature for different solid polymer electrolyte

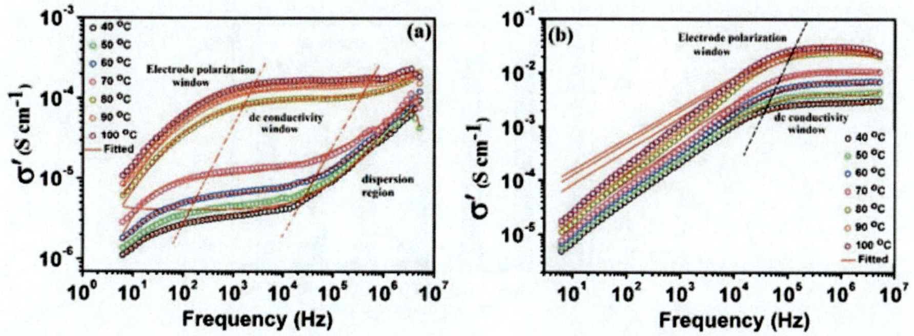

Fig. 4 The frequency and temperature dependence of real part of complex conductivity, **a** pure PEO/PVP, **b** PEO/PVP with salt O/Na = 14. The solid red line is fitted plot

films. The investigation of all the sample shows the increase of ac conductivity with an increase of temperature and frequency. The complex conductivity plot is divided into three regions depending on the frequency.

The low-frequency region is associated with the electrode polarization region, an intermediate frequency region associated with dc conductivity, and high-frequency dispersion region [21–24]. The third region is associated with a short-range path for the ion. It is noticed that with increase of temperature, hopping frequency shifts towards the high-frequency region and indicates the faster ion migration. There are two phenomena occurring at the high-frequency window (i) successful hopping; coordinating site relaxation when ion achieves new coordination site, and (ii) unsuccessful hopping; ion jumps back to the previous site coordinating [10]. The solid red line is the best-fitted plot in the whole frequency range (1 Hz–1 MHz).

Figure 5a, b shows the frequency-dependent imaginary part of conductivity at different temperature for the blend polymer and blend polymer salt complex. With the decrease of frequency, the imaginary part of conductivity decreases (right to left in the plot) and followed by a minimum in the plot corresponding to a frequency known as onset frequency (ω_{on}). At this frequency, the polarization built upstarts and with a further decrease of frequency maximum frequency (ω_{max}). At this frequency maximum built up of polarization occurs. With the further decrease of frequency conductivity again decreases for both polymeric systems. It is important to note

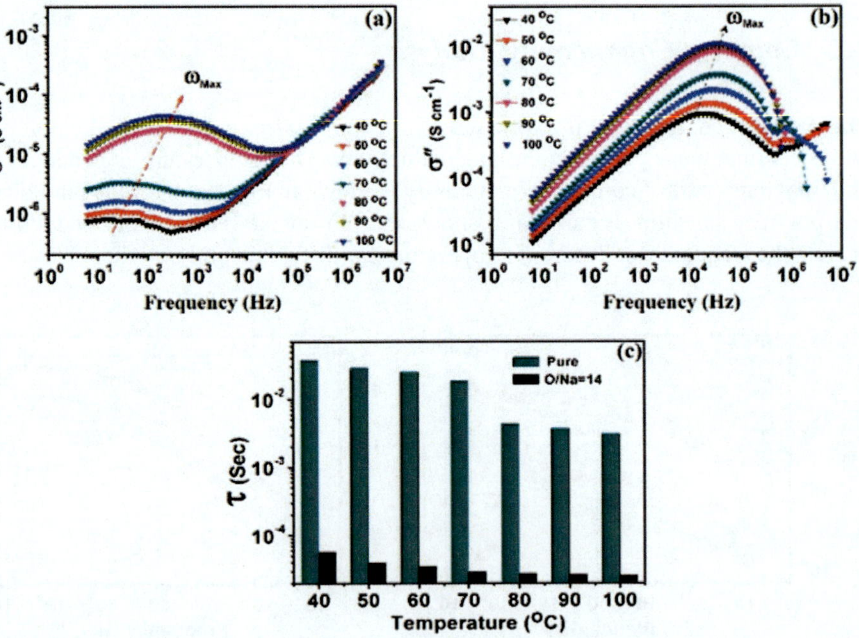

Fig. 5 The frequency and temperature-dependent imaginary part of complex conductivity, **a** pure PEO/PVP, **b** PEO/PVP with salt O/Na = 14, **c** temperature-dependence of relaxation time

that both, ω_{max} and ω_{on} shift towards the high-frequency region with an increase in the temperature. It suggests the enhanced ion migration and is examined by calculating the relaxation time [10, 25, 26]. It is concluded that the mainly effective role of temperature, as enhancement the conductivity then all samples for temperature-dependent in Fig. 5a, b.

Figure 5c shows the variation of relaxation time for different temperature compared with the pure polymer and polymer-salt matrix. It may be noted that with an increase in temperature lowering of relaxation time is observed for both pure blend polymer and polymer-salt matrix. Here, the polymer salt matrix shows the lowest relaxation time as compared to the pure blend and it indicates the enhanced dc conductivity on the addition of salt. The relaxation time decreases from 57 s (at 40 °C) to 25 s (at 100 °C). It indicates the enhanced ion dynamics and faster segmental motion of the polymer chain. The ionic conductivity is inversely proportional to the relaxation time, so lowering of relaxation time suggests the enhanced dc conductivity and is in correlation with the complex permittivity results [27, 28]. The lowering of the relaxation time promotes fast ion migration, hence the enhancement of conductivity.

3.3 Ion Transport Mechanism

Figure 6a shows the general interaction mechanism of the polymer and salt matrix. The cation migrates via the coordinating sites provided by the polymer chains. While anion due to large size remains in the immobilized state and interacts with polymer backbone. Figure 6b shows the effect of the temperature on the polymer salt matrix. Initially, the polymer matrix in the semi-crystalline state and addition of salt disrupts the polymer chain arrangement. When the temperature is increased then it further reduces the crystalline arrangement and disorder in the matrix increases. This results in faster ion migration. Initially, at low-temperature hopping process dominates and short-range ion migration takes place. While, with an increase of temperature, thermal

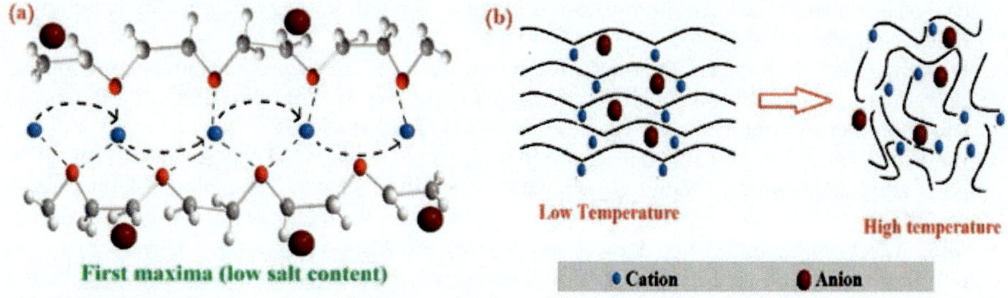

Fig. 6 **a** Ion migration in the polymer matrix, and **b** crystalline to amorphous transition with temperature

activation of charge carriers permits the long-range conduction path. The simultaneous presence of both phenomena results in the overall enhanced conductivity. This is also supported by the lowering of the relaxation time with the increase of temperature. In brief, it may be concluded that the temperature effectively enhances the ion dynamics in the blend solid polymeric matrix.

4 Conclusions

A novel blend solid polymer electrolyte has been synthesized using the blended polymer PEO and PVP and sodium salt ($NaNO_3$) via standard solution cast technique. The fitting result of complex dielectric permittivity and complex conductivity are in close agreement with the experimental data. The effect of variation of temperature on the dielectric properties of solid polymer electrolyte film has been investigated. The complex dielectric permittivity analysis evidences the enhancement of the dielectric constant with the increase of temperature and indicates the enhancement in the number of free charge carriers. It evidences the faster ion migration associated with the fast segmental motion of the polymer chain. The dielectric constant, dielectric strength and ac conductivity increase with the temperature while the relaxation time decrease. The ac conductivity increase with the increase in temperature suggests the faster ion migration. The ion transport mechanism has been proposed and temperature effectively enhances the ion dynamics of the solid polymer electrolyte film. It may be concluded that the present BSPE has the potential to be used as an electrolyte for application in sodium-ion batteries (SIB).

References

1. Arya, A., Sharma, A.L.: Insights into the use of polyethylene oxide in energy storage/conversion devices: a critical review. J. Phys. D Appl. Phys. **50**, 443002 (2017)
2. Pritam, S., Arya, A., Sharma, A.L.: Dielectric relaxations and transport properties parameter analysis of novel blended solid polymer electrolyte for sodium-ion rechargeable batteries. J. Mater. Sci. **54**(9), 7131–7155 (2019)
3. Koduru, H.K., Marino, L., Scarpelli, F., Petrov, A.G., Marinov, Y.G., Hadjichristov, G.B., Scaramuzza, N.: Structural and dielectric properties of $NaIO_4$–Complexed PEO/PVP blended solid polymer electrolytes. Curr. Appl. Phys. **17**, 1518–1531 (2017)
4. Barbosa, J.C., Dias, J.P., Lanceros-Méndez, S., Costa, C.M.: Recent advances in poly(Vinylidene fluoride) and its copolymers for lithium-ion battery separators. Membranes **8**, 45 (2018)
5. Sim, L.N., Sentanin, F.C., Pawlicka, A., Yahya, R., Arof, A.K.: Development of polyacrylonitrile-based polymer electrolytes incorporated with lithium bis (trifluoromethane) sulfonimide for application in an electrochromic device. Electrochim. Acta **229**, 22–30 (2017)
6. Arya, A., Sharma, A.L.: A Glimpse on All-Solid-State Li-Ion Battery (ASSLIBs) performance based on novel solid polymer electrolytes: a topical review. J. Mater. Sci. **55**, 6242–6304 (2020)

7. Yue, L., Ma, J., Zhang, J., Zhao, J., Dong, S., Liu, Z., Chen, L.: All solid-state polymer electrolytes for high-performance lithium ion batteries. Energy Storage Mater. **5**, 139–164 (2016)
8. Chapi, S., Raghu, S., Devendrappa, H.: Enhanced electrochemical, structural, optical, thermal stability and ionic conductivity of (PEO/PVP) polymer blend electrolyte for electrochemical applications. Ionics **22**, 803–814 (2016)
9. Kumar, K., Ravi, M., Pavani, Y., Bhavani, S., Sharma, A.K., VVR, N.R.: Electrical conduction mechanism in NaCl complexed PEO/PVP polymer blend electrolytes. J. Non-Crystall. Solids **358**, 3205–3211 (2012)
10. Arya, A., Sharma, A.L.: Structural, electrical properties and dielectric relaxations in Na+-ion-conducting solid polymer electrolyte. J. Phys.: Condens. Matter **30**, 165402 (2018)
11. Roy, A., Dutta, B., Bhattacharya, S.: Correlation of the average hopping length to the ion conductivity and ion diffusivity obtained from the space charge polarization in solid polymer electrolytes. RSC Adv. **6**, 65434–65442 (2016)
12. Arya, A., Sharma, A.L.: Tailoring of the structural, morphological, electrochemical, and dielectric properties of solid polymer electrolyte. Ionics **25**, 1617–1632 (2019)
13. Kim, S., Park, S.J.: Preparation and ion-conducting behaviors of poly (ethylene oxide)-composite electrolytes containing lithium montmorillonite. Solid State Ionics **178**, 973–979 (2007)
14. Klongkan, S., Pumchusak, J.: Effects of nano alumina and plasticizers on morphology, ionic conductivity, thermal and mechanical properties of PEO-LiCF3SO_3 solid polymer electrolyte. Electrochim. Acta **161**, 171–176 (2015)
15. Pradhan, D.K., Karan, N.K., Thomas, R., Katiyar, R.S.: Coupling of conductivity to the relaxation process in polymer electrolytes. Mater. Chem. Phys. **147**, 1016–1021 (2014)
16. Arya, A., Sharma, A.L.: Effect of salt concentration on dielectric properties of Li-ion conducting blend polymer electrolytes. J. Mater. Sci.: Mater. Electron. **29**, 17903–17920 (2018)
17. Kesavan, K., Mathew, C.M., Rajendran, S.: Lithium ion conduction and ion-polymer interaction in poly (vinyl pyrrolidone) based electrolytes blended with different plasticizers. Chin. Chem. Lett. **25**, 1428–1434 (2014)
18. Tang, R., Jiang, C., Qian, W., Jian, J., Zhang, X., Wang, H., Yang, H.: Dielectric relaxation, resonance and scaling behaviors in $Sr_3Co_2Fe_{24}O_{41}$ hexaferrite. Sci. Rep. **5**, 13645 (2015)
19. Sharma, A.L., Thakur, A.K.: AC conductivity and relaxation behavior in ion conducting polymer nanocomposite. Ionics **17**, 135–143 (2011)
20. Arya, A., Sadiq, M., Sharma, A.L.: Salt concentration and temperature dependent dielectric properties of blend solid polymer electrolyte complexed with NaPF6. Mater Today: Proc. **12**(3), 554–564 (2019)
21. Arunkumar, R., Babu, R.S., Rani, M.U., Rajendran, S.: Influence of plasticizer on ionic conductivity of PVC–PBMA polymer electrolytes. Ionics **23**, 3097–3109 (2017)
22. Sharma, A.L., Thakur, A.K.: Relaxation behavior in clay-reinforced polymer nanocomposites. Ionics **21**, 1561–1575 (2015)
23. Kumar, P.S., Sakunthala, A., Govindan, K., Reddy, M.V., Prabu, M.: Single crystalline TiO_2 nanorods as effective fillers for lithium ion conducting PVdF-HFP based composite polymer electrolytes. RSC Advances **6**, 91711–91719 (2016)
24. Arya, A., Nilesh, S.G., Sharma, A.L.: Impact of shape (Nanofiller vs. Nanorod) of TiO_2 nanoparticle on free standing solid polymeric separator for energy storage/conversion devices. J. Appl. Polymer Sci. **136**, 47310 (2018)
25. Choudhary, S., Sengwa, R.J.: Structural and dielectric studies of amorphous and semicrystalline polymers blend-based nanocomposite electrolytes. J. Appl. Polym. Sci. **15**, 132 (2015)
26. Dhatarwal, P., Sengwa, R.J., Choudhary, S.: Effectively improved ionic conductivity of montmorillonite clay nanoplatelets incorporated nanocomposite solid polymer electrolytes for lithium ion-conducting devices. SN Appl. Sci. **1**, 112 (2019)

27. Nath, A.K., Kumar, A.: Scaling of AC conductivity, electrochemical and thermal properties of ionic liquid based polymer nanocomposite electrolytes. Electrochim. Acta **129**, 177–186 (2014)
28. Cao, W., Gerhardt, R.: Calculation of various relaxation times and conductivity for a single dielectric relaxation process. Solid State Ionics **42**, 213–221 (1990)

In Situ X-Ray Diffraction and Alkali Ion (A = Li, Na, K) Intercalation Behavior of $Na_2FeP_2O_7$ Pyrophosphate

Ritambhara Gond, Shashwat Singh, Sai Pranav Vanam, Valerie Pralong, and Prabeer Barpanda

Abstract Among polyanionic materials, $Na_2FeP_2O_7$ system has attracted much attention towards large-scale economic sodium-ion battery applications. High thermal stability, facile Na-ion migration, and high operational stability has made this system a suitable subject to explore Li-ion and K-ion intercalation into the pyrophosphate host matrix. Pyrophosphate $(P_2O_7)^{4-}$ anion offers better safety than other phosphate-based cathodes. Further, Fe-based systems are less hazardous and safer electrode materials and can offer a new class of cathode materials for Li-ion and K-ion battery applications. Excellent electrochemistry of triclinic $Na_2FeP_2O_7$ structure (space group: $P\bar{1}$) has been studied using in situ XRD for few cycles tested against Na-ion intercalation (vs. Na/Na^+), which clearly supports and demonstrates the expansion and contraction of the structure during Fe^{3+}/Fe^{2+} redox process.

Keywords Polyanions · Pyrophosphate · In situ XRD

1 Introduction

To address the challenge of energy storage, utilization of renewable energy resources as rechargeable batteries has been opted long before. Rechargeable batteries have shown its crucial role in transition devices, portable electronic devices, and electric vehicles. Predominantly rechargeable Li-ion batteries (LIB) rule the sector of electrochemical energy storage devices [1]. Additionally, rechargeable Na-ion batteries (SIB) have also shown their importance in stationary grid storage devices for very large-scale applications [2]. Along with the rising energy demand and to support sustainable energy storage especially for large-scale, secondary batteries are required to be complemented with the earth-abundant potassium ion based chemistry from the

R. Gond (✉) · S. Singh · S. P. Vanam · P. Barpanda
Faraday Materials Laboratory, Materials Research Center, Indian Institute of Science, Bangalore 560012, India
e-mail: ritambharag@iisc.ac.in

V. Pralong
ENSICAEN, UNICAEN, CNRS, CRISMAT, Normandie University, 14000 Caen, France

Y. Sharma et al. (eds.), *Recent Research Trends in Energy Storage Devices*,
https://doi.org/10.1007/978-981-15-6394-2_15

same group (group I). Potassium ion batteries (KIB) have attracted much attention recently and various compounds are proposed and investigated [3].

In the race to develop novel positive cathodes, various compounds have already been proposed and investigated from "pyrophosphate-based" polyanionic materials [4]. The present article offers an electrochemical investigation of various group I elements (cations intercalation) against one particular sodium-based pyrophosphate system. Based on earth-abundant elements, $Na_2FeP_2O_7$ can form an economic cathode for grid storage. Thus, in this work, we targeted $Na_2FeP_2O_7$ to intercalate Li-ion as well as K-ion. $Na_2FeP_2O_7$ demonstrates three-dimensional channels running along [100], [-110] and [01-1] directions [5]. It offers high thermal stability and zero thermal decomposition or oxygen evolution during charge–discharge which has established this pyrophosphate as a safe cathode candidate. These remarkable properties were further characterized by in situ XRD technique applied over $Na_2FeP_2O_7$ during the galvanostatic charge–discharge experiment for SIB analogue. The underlying redox mechanism was investigated on a custom-made in situ cell tested for Na-ion intercalation which will help us understand the structural stability due to the robust framework built of $(P_2O_7)^{-4}$ units.

2 Experimental

2.1 *Material Synthesis*

Pristine $Na_2FeP_2O_7$ was synthesized by eco-efficient sustainable synthesis which is a two-step splash combustion synthesis [6]. Splash combustion synthesis method involved the dissolution of a stoichiometric mixture of $NaH_2PO_4{\bullet}H_2O$ (Merck, $\geq$98%), $Fe(NO_3)_3{\bullet}9H_2O$ (Merck, $\geq$98%) and $C_6H_8O_6$ (SDFCL, $\geq$99.5%), where first, low-cost Fe^{3+} precursor [$Fe(NO_3)_3{\bullet}9H_2O$] was dissolved in distilled water. Then a pinch of ascorbic acid ($C_6H_8O_6$) was added to the solution in order to convert the Fe^{3+} into Fe^{2+} state which was evident by the change in color from brown to colorless/light green solution. Further other precursors were added to form a homogeneous mixture of the precursor solution. In the end, 1 mol of ascorbic acid (fuel) was added to propel the combustion reaction. Initially, the reaction mixture was first subjected to heat treatment for the removal of excess water at 120 °C. Later, temperature was raised to 300 °C to propel exothermic combustion reaction leading to the formation of intermediate combustion complex. This complex was ground, pelletized and annealed at 600 °C for 6 h in a tubular furnace with steady Ar flow.

2.2 *Material Characterization*

The structure of as-synthesized pyrophosphate was verified by powder X-ray diffraction (XRD) using PANalytical X'Pert Pro diffractometer with Cu-Kα source (λ_1= 1.5405 Å, λ_2= 1.5443 Å) operating at 40 kV/30 mA. Using GSAS program, Rietveld refinement was done with the EXPGUI front-end software [7]. $Na_2FeP_2O_7$ crystal structure was illustrated using the VESTA software [8]. Scanning electron microscopy (FEI Inspect F50 microscope) and transmission electron microscopy (FEI Tecnai F 30 STwin) were performed on the powder sample to get the morphological information, for which SEM and TEM were operated at 10 kV and 200 kV, respectively.

3 Result and Discussion

3.1 *Structure and Electrochemical Activity*

$Na_2FeP_2O_7$ exhibited triclinic (space group: $P\bar{1}$) structure, where each corner of FeO_6 octahedra is bridged to $(P_2O_7)^{-4}$ units. Corner shared FeO_6 creates Fe_2O_{11} dimers, which are interconnected by corner-sharing as well as edge-sharing with $(P_2O_7)^{-4}$ groups. Corresponding XRD pattern is shown in Fig. 1a with the VESTA image of $Na_2FeP_2O_7$ along *bc*-plane. The staggered connections of FeO_6 and PO_4

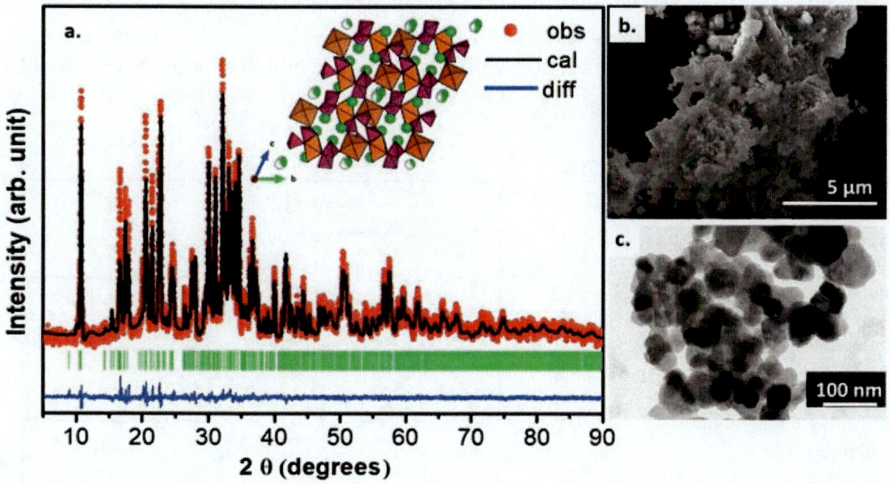

Fig. 1 **a** Rietveld refined XRD patterns of $Na_2FeP_2O_7$ prepared by solution combustion route. **b** SEM and **c** TEM image of $Na_2FeP_2O_7$ particle

give rise to large tunnels for Na atom along [011] direction. Crystallographic positioning of Na in this pyrophosphate is quite interesting, as this system has six crystallographic sites for Na atom with partial occupancies for three Na atom into the structure. Literature suggests the partial occupancies and low activation energies for long-range diffusion in all crystallographic directions in $Na_2FeP_2O_7$ suggesting three-dimensional (3D) Na-ion diffusion [9]. Due to which this material as cathode has shown remarkable galvanostatic charge–discharge behavior against sodium half-cell architecture. Fe has two crystallographic positions with identical occupancies.

Seeing beautiful 3D nature for Na-ion migration lithium-ion and potassium-ion intercalation were tested in lithium and potassium ion half-cell architecture, respectively. Sodium half cells were also made for in situ XRD experiments to understand the planes involved during sodium-ion migration. Morphology of the synthesized powder (Fig. 1b,c) suggests no further particle downsizing is required before electrode preparation. Thus 80 ($Na_2FeP_2O_7$):10 (C): 10 (PVDF) is the compositional ratio taken for electrode preparation using NMP solvent. Coin cells (CR2032) were made against lithium and potassium ion half-cell architecture. For Li-ion half-cell 1 M $LiPF_6$ in EC:PC:DMC (1:1:3 v/v) as electrolyte was used. While for K-ion half-cell 0.8 M KPF_6 in EC:DEC (1:1v/v) solution was used as electrolyte. For both cases, glass fibre was used as a separator. To perform in situ XRD only 80 wt% of $Na_2FeP_2O_7$ was added with 20 wt% of Carbon super P. 1 M $NaClO_4$ dissolved in propylene carbonate (PC) was used as an electrolyte in a custom-made Na half-cell architecture.

During the galvanostatic charge–discharge (tested at C/20 rate) for the first cycle against Li and K half-cell, sodium extraction from the structure has followed different paths individually versus Li as well as versus K which is clear from Fig. 2. This may be possible due to different electrolytes used. Once pristine $Na_2Fe^{2+}P_2O_7$ is subjected to first charge, i.e., after de-sodiation it will give rise to $Na_{2-x}Fe^{3+}P_2O_7$ end member involving a Fe^{3+}/Fe^{2+} redox reaction. Further during the first discharge insertion of Li and K-ion will give rise to $Li_xNa_{2-x}Fe^{2+}P_2O_7$ and $K_xNa_{2-x}Fe^{2+}P_2O_7$, respectively.

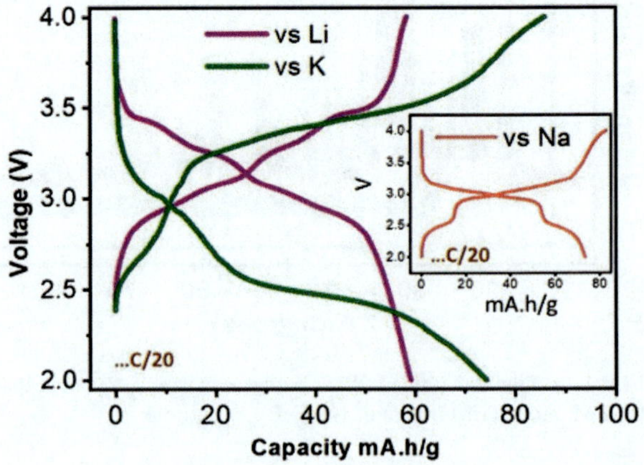

Fig. 2 Galvanostatic charge–discharge voltage profiles for Li-ion insertion (~60 mAh/g) and K-ion insertion (~77 mAh/g) in $Na_2FeP_2O_7$ shown in purple and green color, respectively. Inset Na-ion insertion in $Na_2FeP_2O_7$ as reference delivering capacity of ~78 mAh/g

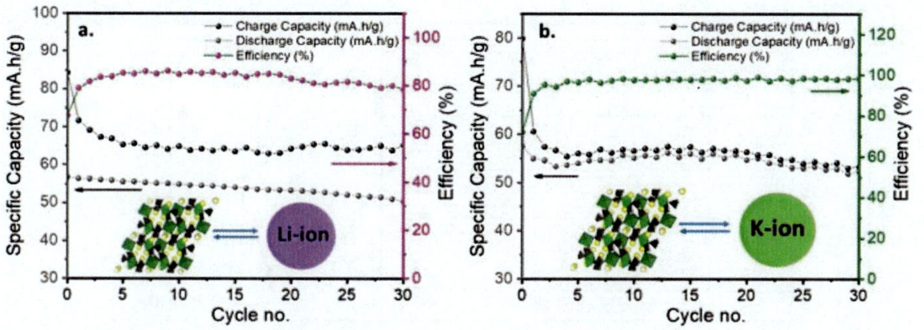

Fig. 3 Cycling stability and galvanostatic charge–discharge capacity of **a** Li-ion insertion/extraction and **b** K-ion insertion/extraction are shown for $Na_2FeP_2O_7$ against percentage coulombic efficiency

$Na_2FeP_2O_7$ against Li-ion half-cell delivered 60 mAh/g with Fe^{3+}/Fe^{2+} redox reaction at 3.1 V. Similarly, in K-ion half-cell architecture this pyrophosphate delivered 77 mAh/g of specific capacity with an average working potential of approximately 2.9 V same as it exhibits with Na-intercalation. During K-ion insertion huge polarization observed which is reflecting further optimization need to be done with respect to the electrolyte used and opens an area for a researcher to minimize this polarization.

Figure 3 shows the galvanostatic cycling stability of Li and K half-cell at C/20 rate. Li or K insertion/extraction has been shown in Fig. 3 against the host framework for better understanding. Efficiency for the LIB analogue is found to be 80% with capacity retention up to 30 cycles. Similarly, for KIB analogue the efficiency was observed to be 98% after few cycles. For Li half-cell charge capacities were observed to be higher than discharge capacity, however, for K half-cell these capacities were found to be similar due to size issue. It suggests K-ion insertion/extraction into the $Na_2FeP_2O_7$ is efficient than Li-ion. However, there has been an increase in the average potential for Fe^{3+}/Fe^{2+} redox with Li half-cell.

3.2 In Situ *Electrochemistry in Na Half-Cell*

In situ XRD is a powerful tool to gauze the crystallographic-electrochemical information of electrode material [10]. In the present case during the redox process, $Na_2FeP_2O_7$ goes through various mixed-valence intermediate compositions [$Na_{2-x}Fe^{2+/3+}P_2O_7$, $x = 0–1$] with steady structural ordering/reorientation during galvanostatic charge–discharge in Na half-cell configuration. These intermediate stages are traced with a combination of electrochemical techniques and in situ X-ray diffraction technology which is shown in Fig. 4.

The results obtained from in situ XRD pattern recorded during extraction (in red color) and insertion (in blue color) of Na from the host material are very clear

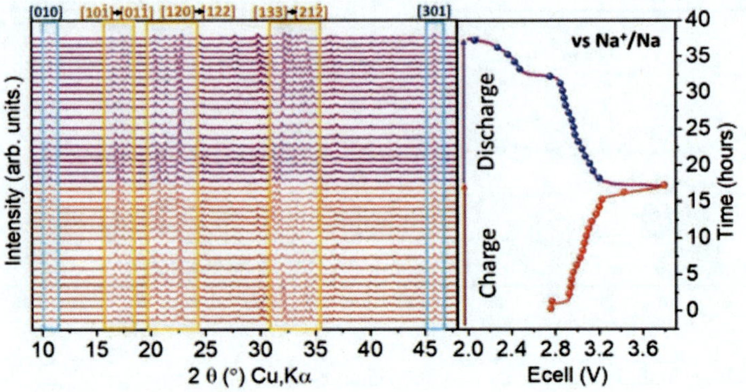

Fig. 4 In-situ XRD pattern with miller indices involved during charge–discharge of $Na_2FeP_2O_7$ at C/25 rate from 2 to 3.7 V (left) where patterns in red color shows Na-ion de-intercalation from the host matric during the first charge and the patterns in blue color shows Na-ion insertion during first discharge presented as Ecell versus time (right)

from the phases which actually tells what are the crystal planes involved for Na-ion migration. Diffraction patterns were recorded for 3624 s at an interval of 2 s in the angular domain of 9°–60°. First peak coming at 10.78° 2θ value corresponds to [010], and peak at 45.89° 2θ value of [301] plane does not show any change during the sodium intercalation/de-intercalation from $Na_2FeP_2O_7$. Thus, it clearly depicts that these two planes are not involved in sodium extraction and insertion shown with cyan-colored box in Fig. 4. The planes involved during migration of charge carrier are shown with yellow boxes in Fig. 4 which are in the 2θ range of 15°–18°, 20°–25°, and 30°–35°. [011] are the planes where tunnel exists to accommodate Na atom thus peak changes at 16.4° and 16.7° 2θ value corresponds to [10-1] and [01-1] plane reveals the extraction and insertion of Na-ion. Similarly, planes at 22.5° and 22.8° are [120] and [122] which are submerging into one peak at 3 V potential and again splitting into two peaks at other voltages. During Na-ion migration stronger repulsive interaction exists between electronegative oxygens causing volume expansion which can be verified by shift in the 2θ value towards lower angle at 32° corresponds to [133] plane. It can be visualized by an increase in d spacing due to the expansion of FeO_6 octahedra accompanied by Na atom tunnel. However, peak at 34° move towards higher 2θ angle of [21-2] miller plane reveals the lattice contraction which could be the space between pyrophosphate unit and FeO_6 octahedra lying adjacent to tunnel resulted from expansion of two neighboring FeO_6 in front of the tunnel. A close look in these diffraction peaks shows reversible intensities during charge–discharge. Thus, based upon in situ operando diffraction experiments $Na_2FeP_2O_7$ is found to be very stable due to strong framework built by pyrophosphate units and can easily intercalate Li-ion as well as K-ion into the structure without any structure collapse.

4 Conclusion

In summary, we have reported the Li-ion and K-ion intercalation in triclinic $Na_2FeP_2O_7$. With Na-ion this system delivered ~78 mAh/g with Fe^{3+}/Fe^{2+} redox potential of 2.9 V. However, with Li-ion and K-ion the capacity was found to be ~60 mAh/g and ~77 mAh/g, respectively at an average potential of 3.1 and 2.9 V for Li-ion and K-ion analogue, respectively. In situ XRD operated against Na-ion battery encouraged intercalation mechanism with huge stability.

Acknowledgements The authors thank Shell Technology Centre (STC) Bangalore for financial support. Also, RG acknowledges the University Grants Commission (UGC) for financial support. SS and SPV thanks to Ministry of Human Resource Development (MHRD) for financial support.

References

1. Scrosati, B., Garche, J.: Lithium batteries: status, prospects and future. J. Power Sources **195**(9), 2419–2430 (2010)
2. Hwang, J.-Y., Myung, S.-T., Sun, Y.-K.: Sodium-ion batteries: present and future. Chem. Soc. Rev. **46**(12), 3529–3614 (2017)
3. Hosaka, T., Shimamura, T., Kubota, K., Komaba, K.: Polyanionic compounds for potassium-ion batteries. Chem. Rec. **19**(4), 735–7451 (2019)
4. Barpanda, P., Nishimura, S., Yamada, A.: High-voltage pyrophosphate cathodes. Adv. Energy Mater. **2**(7), 841–859 (2012)
5. Barpanda, P., Liu, G., Ling, C.D., Tamaru, M., Avdeev, M., Chung, S.-C., Yamada, Y., Yamada, Y.: $Na_2FeP_2O_7$: a safe cathode for rechargeable sodium-ion batteries. Chem. Mater. **25**(17), 3480–3487 (2013)
6. Barpanda, P., Ye, T., Chung, S.-C., Yamada, Y., Nishimura, S., Yamada, A.: Eco-efficient splash combustion synthesis of nanoscale pyrophosphate ($Li_2FeP_2O_7$) positive-electrode using Fe(III) precursors. J. Mater. Chem. **22**(27), 13455–13459 (2012)
7. Larson, A.C., Dreele, R.B.: General Structure Analysis System (GSAS). Los Alamos National Laboratory Report. LAUR 86–748, Los Alamos National Laboratory, NM (1994)
8. Momma, K., Izumi, F.: VESTA 3 for three-dimensional visualization of crystal, volumetric and morphology data. J. Appl. Crystallogr. **44**(6), 1272–1276 (2011)
9. Clark, J.M., Barpanda, P., Yamada, A., Islam, M.S.: Sodium-ion battery cathodes $Na_2FeP_2O_7$ and $Na_2MnP_2O_7$: diffusion behaviour for high rate performance. J. Mater. Chem. A **2**(30), 11807–11812 (2014)
10. Sharma, N., Pang, W.K., Guo, Z., Peterson, V.K.: In situ powder diffraction studies of electrode materials in rechargeable batteries. Chemsuschem **8**(17), 2826–2853 (2015)

Fabrication and Electrochemical Performance of Low-Cost Soluble Lead Redox Flow Battery Using Two Different Carbon Electrodes

Shivangi Kosta, R. Sneha, and Kuldeep Rana

Abstract The existing approach focuses on the fabrication of simple cost-effective soluble lead redox flow batteries using a different carbon-based electrode. Two different soluble lead redox flow cells with reduced graphene oxide (rGO) and ultra-thin graphite (UG) as electrode material are fabricated and their electrochemical performance has been studied individually. The different carbon material—rGO and UG were synthesized using wet chemical and chemical vapor deposition (CVD) method respectively. The synthesized carbon nanostructures were characterized by X-ray diffraction, Raman spectroscopy, and Scanning Electron Microscopy. Electrolyte consists of solution of lead (II) methanesulfonate and methanesulfonic acid without using any additive which has been circulated in cell through the pump at a flow rate of 90 ml min^{-1}. The electrochemical performance of two different carbon-based SLRFBs single cell has been studied and compared. The average discharge capacity of rGO-SLRFB is around 167 mAh with an average efficiency of 92% has been observed. The rGO-SLRFB cell is tested for 20 charge/discharge cycles at a constant current of 0.5 A. However, in the case of UG-SLRFB the decay in capacity was observed within initial cycles and the failure analysis of the same has been studied in current work.

Keywords Energy storage · Electrode fabrication · Soluble lead redox flow batteries

1 Introduction

Redox flow batteries (RFB) recently have been extensively studied for large scale energy storage applications. Various chemistries of flow batteries such as vanadium, Fe-Cr, Fe-V, Zn-Br, Br-polysulphide are existing today, however, among all other technologies, vanadium redox system is more matured and developed system [1]. All the flow battery systems employ an ion-exchange membrane, which is expensive

S. Kosta · R. Sneha · K. Rana (✉)
Electrical Appliances Technology Division, Central Power Research Institute, Bangalore, India
e-mail: kuldeeprana@cpri.in

Y. Sharma et al. (eds.), *Recent Research Trends in Energy Storage Devices*,
https://doi.org/10.1007/978-981-15-6394-2_16

and increases the complexity associated with unwanted ion transport through this membrane. Further different electrolytes being used for the anode and cathode as anolyte and catholyte, respectively, leading to an increase in the complexity of the system. Therefore, further modifications required to make them commercially viable by lowering the cost by simple design.

The soluble lead redox flow battery (SLRFB) system has shown much promise in reducing the cost and complexity. Single electrolyte, single pump, and absence of membrane reduces the cost of the system and make it less complicated as compared with other flow battery systems. It can also be operated without any separable membrane leading to a decrease in the cost as well as the usage of cell components; overall simplifies the cell design. The performance of the cell is completely based on the fabrication and structure of the electrodes and the electrolyte composition. The reaction site and the current collector determine battery power whereas volume of the electrolyte tank gives battery energy.

The reason why soluble lead redox flow batteries are uniquely opted using single electrolyte is because lead (II) ion is both oxidized and reduced upon charging and discharging of the cell. The electrolyte of SLRFB consists of high concentrated lead methanesulfonate (LMS) and methanesulfonic acid (MSA) circulated between the electrodes by pumping it. During the charging process, the ions Pb^{2+} is oxidized so as to form a deposit of lead dioxide (PbO_2) on the positive electrode and reduced in order to form a deposit of lead metal on the negative electrode. During the discharge process, these deposits are re-dissolved as ions Pb^{2+} in the electrolyte. These processes are summarized using the following reaction:

$$\text{Negative: } Pb^{2+} + 2e^- \underset{\text{DIscharge}}{\overset{\text{Charge}}{\rightleftharpoons}} Pb \qquad E^0 = -0.125\ V$$

$$\text{Positive: } Pb^{2+} + H_2O \underset{\text{DIscharge}}{\overset{\text{Charge}}{\rightleftharpoons}} PbO_2 + 4H^+ + 2e^- \quad E^0 = 1.468\ V$$

$$\text{Overall: } Pb^{2+} + 2H_2O \underset{\text{DIscharge}}{\overset{\text{Charge}}{\rightleftharpoons}} Pb + PbO_2 + 4H^+ \quad E^0 = 1.593\ V$$

Previously, SLRFB was fabricated using perchloric, fluorosilicic, or fluoroboric acid electrolytes that have soluble lead salt [2–5]. However, these were primary batteries fabricated using electrochemical plating or pasting of active material onto the current collectors. These batteries can produce a high current density for a short period of time and are often targeted to small-scale applications; such devices usually return relatively low-voltage efficiency. Several patents [6–8] have described secondary batteries where soluble lead has been introduced as electrolytes. These tend to be small box/button cells with electrolytes typically comprising of perchloric, hexafluorosilicic or fluoroboric acid. The kinetics of the Pb^{2+}/PbO_2 couple were noticeably slower than those of the Pb/Pb^{2+} couple and some inefficiency was introduced due to the lead dioxide electrode and earlier used electrolyte. This inefficiency leads to degradation in the performance of the cell.

The main cause for the failure of soluble lead redox flow batteries is the deposition of lead and lead oxide and the formation of dendrites over the surface of the electrodes, which cause the short circuit in battery and hence degradation in capacity with cycle. Therefore, in order to enhance the performance and solve the dendrite problem, several efforts have been made such as introducing the use of additives in electrolyte which are costly and hence are not feasible for lowering the cost of battery. Henk et al. [8] made use of lead methanesulfonate as an additive to the predominantly lead silicofluoride electrolyte. A series of papers by Pletcher et al. [9–14] has described the electrochemistry and performance of a small 2 and 8 cm^2 flow cell based on an electrolyte containing solvated lead (II) ions in aqueous methanesulfonic acid. Collins et al. [15] fabricated larger cells with active electrode areas of 64 cm^2 and 100 cm^2 and its electrochemical performance was tested using hexadecylytrimethylamine as an additive in the electrolyte, the efficiency observed was less than 90%. Banerjee et al. [16] introduced corrugated graphite sheet and reticulated vitreous carbon as positive and negative current collectors for soluble lead redox flow battery systems. The test was performed for 200 cycles and achieved a faradaic efficiency of 95% with a discharge capacity of 114 $mAhg^{-1}$ using lignosulfonic acid as an additive in the electrolyte. Further, to improve the cycle life and dissolution of lead oxide during the discharge reaction Wallis et al. [17] incorporated Bi^{3+} [7]ions; the bismuth ions cause a shift in the reaction mechanism and kinetics of lead dioxide dissolution during the discharge reaction. Lin et al. [18] added sodium ethanoate in the electrolyte to increase the lifespan of SLRFB up to 500 cycles with 65% of energy efficiency. Various efforts have been made to improve the performance of SLRFB by modifying the electrode and electrolyte.

In current work, two different types of carbon nanostructure materials have been synthesized and used as the electrodes materials in SLRFB as nanocarbon materials are rarely been used as an electrode for such a battery system. The electrochemical performance such as charge/discharge, cycle life and efficiency has been studied. Two different types of carbon materials: reduced graphene oxide (rGO) and ultrathin graphite (UG) film on Ni-substrate have been synthesized by using wet chemical and chemical vapor deposition (CVD) respectively. The rGO-electrode was tested by fabricating 100 cm^2 single cell by coating the synthesized materials by adding PVDF conducting binder by a slurry method generally used for fabrication of Li-ion battery electrode fabrication. However, UG on Ni was directly used in small size cell 9 cm^2. After fabrication single cells, both the cells were connected to the electrolyte tank through the pump. The electrochemical studies of both the cell have been studied in constant current mode and compared.

2 Experimental Details

GO was synthesized using improved Hummer's method described previously [19]. First 3 g of graphite and 18 g of $KMnO_4$ were mixed in 500 ml round bottom flask. A 9:1 volume ratio of concentrated H_2SO_4 and H_3PO_4 (360:40 ml) was added to the above mixture that leads to an exothermic reaction thereby increasing the temperature of the reaction mixture up to 40–45 °C. After this, the reaction mixture was further heated to 50 °C and held at this temperature for 6 h with continuous stirring. It was then transferred to an ice bath where 30% of H_2O_2 solution was added to the mixture. Final product was centrifuged (at 4000 rpm for 4 h) and the supernatant was decanted. This was further washed with 200 mL of DI water, 200 mL of 30% HCl and 200 mL of ethanol and finally vacuum dried overnight at room temperature.

GO was reduced using hydrazine hydrate (N_2H_4) as the reducing agent. For this 100 mg of GO was sonicated in 100 mL of DI Water. Then, 1 mL of N_2H_4 was added to the GO aqueous dispersion and heated at 100 °C for 24 h. The obtained rGO material was found precipitated at the bottom of the flask as the black solid. The additive-free ultrathin graphite film over Ni foil was grown using CVD process described previously [20].

The electrolyte used in SLRFB has been synthesized using lead methane sulfonate (LMS, $Pb(SO_3CH_3)_2$, 50% of H_2O), and methane sulfonic acid (MSA, HSO_3CH_3). An aqueous solution of the electrolyte was prepared using deionized (DI) water and molarity concentration was kept at 1 M. The electrolyte was used as synthesized.

2.1 *Fabrication of SLRFBs*

A UG-SLRFB was designed and fabricated using ultrathin graphite film over Nickel foil as an electrode, which also functioned as the current collector. The electrodes were kept in a container filled with an electrolyte consisting of inlet and outlet for continuous flow of electrolyte. The electrolyte was circulated inside the container using a rotor pump (Ravel-Peristaltic Pump) at a flow rate of 20 ml min^{-1}. The graphical representation of the setup has been shown (see Fig. 1a). Finally, the cell was connected to a Battery Life Cycle Tester (Bitrode, LCV16-100-12) and its electrochemical behavior was tested.

The cell consists of electrodes which were 12 cm × 12 cm graphite plates and secured to current collectors (copper plates, 12 cm × 12 cm) with an exposed electrode surface area of 10 cm × 10 cm. The inter-electrode gap of 0.4 cm was established between the electrodes, FRG gaskets were inserted between the frames of the electrodes to seal the cell. The electrolyte frame has holes drilled into the sides of the electrodes which were connected with the inlet and outlet pipe for the continuous flow of the electrolyte. The electrolyte was pumped through the cell at a constant flow rate of 90 mL min^{-1}. The positive and negative electrodes were designed and fabricated using rGO. The slurry of rGO was prepared using conducting binder polyvinylidene difluoride (PVDF) was coated over graphite surface and dried in vacuum oven at 100 °C for 5 h. The schematic of which has been demonstrated (see Fig. 1b).

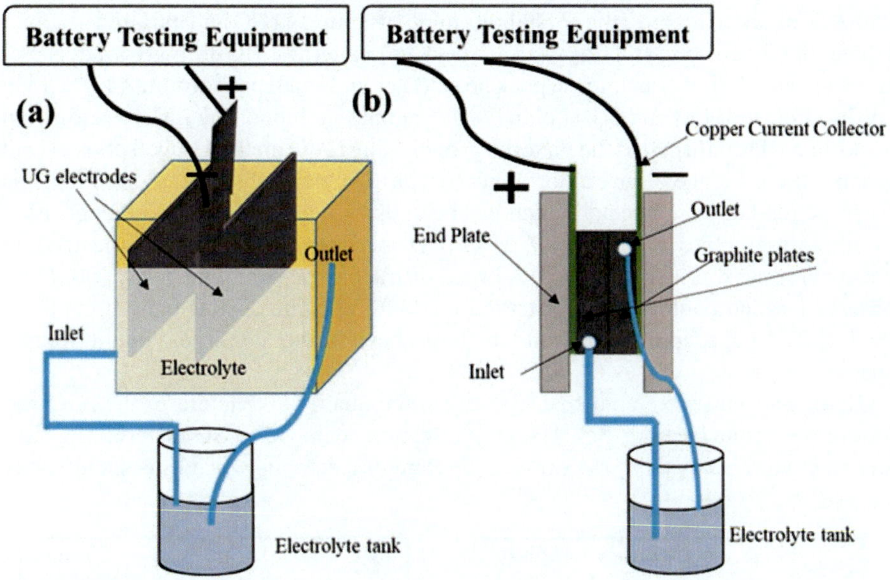

Fig. 1 **a** Schematic representation of ultrathin graphite SLRFB setup. **b** rGO-SLRFB setup

3 Results and Discussion

The Scanning Electron Microscopy (SEM) images of UG and rGO are shown (see Fig. 2). The SEM image of rGO shows layered kind of structure and a random morphology (see Fig. 2a). The microsurface of Ni-substrate is relatively smooth. After the synthesis, the surface of the Ni-substrate was uniformly covered by the ultrathin graphite film [20, 21] (see Fig. 2b). The synthesis of rGO and ultrathin graphite was further confirmed by XRD and Raman analysis.

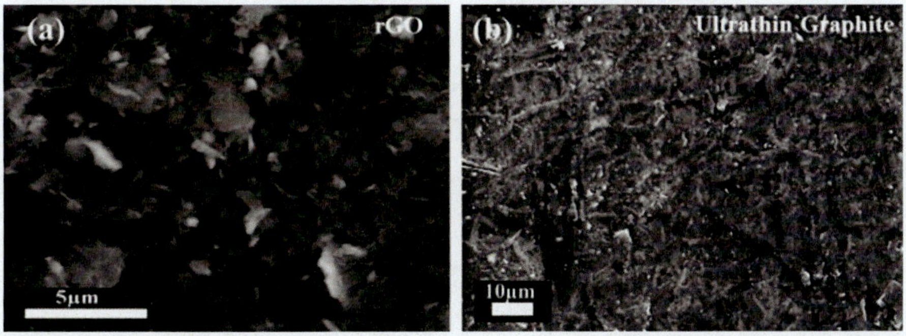

Fig. 2 **a** SEM image of rGO synthesized using chemical route. **b** SEM image of ultrathin graphite grown on Ni-substrate using CVD method

XRD is used to identify the crystallographic structure of rGO and ultrathin graphite grown on Ni-substrate (see Fig. 3a,b). The XRD pattern of UG-film grown over Ni-substrate which shows an intense peak around 26° and small peak around 44.4° which corresponds to [002] and [101] plane of hexagonal-graphite. The intense and sharp peaks in XRD confirm that the material grown using CVD method is well crystallized without the presence of any amorphous carbon. The peak observed around 51.9° is corresponded to the Ni-metal which has been used as a substrate for the deposition of ultrathin graphite materials [20, 21]. XRD spectrum for the rGO synthesized by chemical route was also analyzed. A broad diffraction peak around at 23° [002] was observed which confirms the formation of rGO [22]. Thus, XRD pattern confirms the formation of graphitic material which has been further confirmed by the Raman spectra.

Raman spectra were analyzed to confirm the quality and nature of the rGO and ultrathin graphite (see Fig. 3c). The graphitization degree of a rGO material is generally characterized by the I_D/I_G value in the Raman spectra in which I_G is the intensity

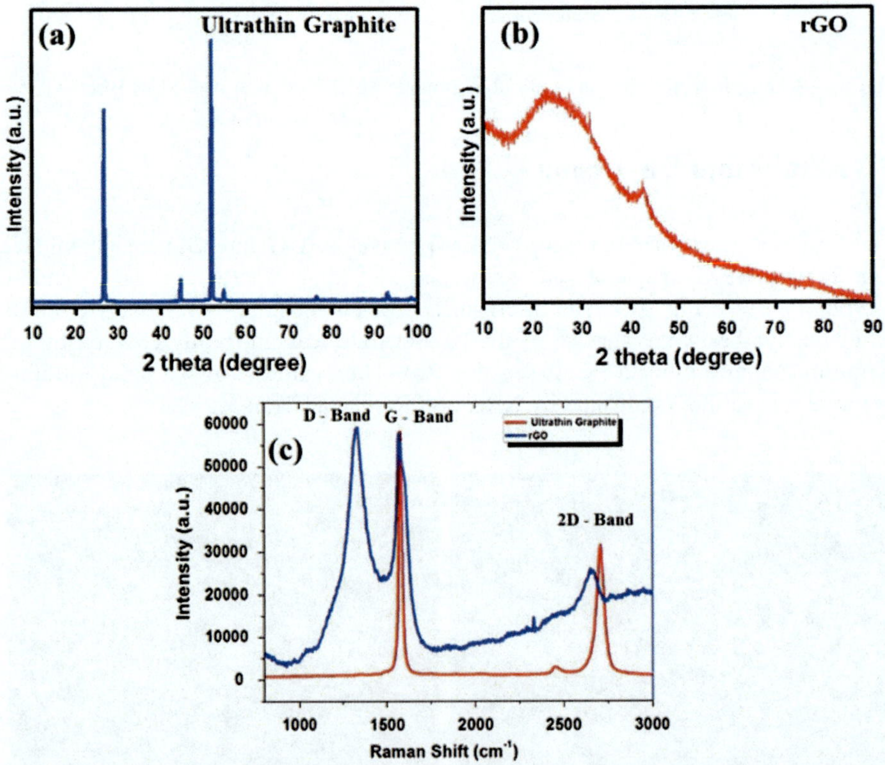

Fig. 3 **a** XRD spectra of ultrathin graphite grown on Ni-substrate using CVD method. **b** XRD spectra of rGO synthesized using chemical route. **c** Raman spectra of rGO and UG

of G-band around 1572 cm^{-1} corresponding to sp^2-bonded carbon atoms in a hexagonal graphitic ring, and I_D is the intensity of D-band around 1335 cm^{-1} corresponding to vibration of sp^3-bonded carbon atoms of defects and disorder, the other peak at 2704 cm^{-1}, corresponds to the 2D-band which confirms the crystalline structure of rGO [22]. The D-band of rGO is more intense as compared to G-band (see Fig. 3c) indicate that rGO has much higher defects and disorder in crystal structure. The resulting spectrum of UG shows two sharp peaks: the peak at 1580 cm^{-1} corresponds to the G-band, which arises from the in-plane vibrations of carbon atoms; the other peak, at 2698 cm^{-1}, corresponds to the 2D-band which confirms that the film is well crystalline in nature without any defects [23].

3.1 Electrochemical Performance

The charge/discharge profiles of a single cell using as synthesized electrode materials UG-SLRFB and rGO-SLRFB (see Fig. 4). These materials have already been demonstrated as the electrode material for lithium-ion batteries [20]. The active area for rGO-SLRFB and UG-SLRFB was 100 cm^2 and 9 cm^2 with an inter-electrode gap of 0.4 cm and 1 cm respectively. The electrolyte consists solution mixture of 1 M $Pb(CH_3SO_3)_2$ + 1 M CH_3SO_3H for both the SLRFB single cell. The flow rate of the electrolyte was controlled by a peristaltic pump and was fixed at 90 mL min^{-1} and 20 mL min^{-1} for rGO-SLRFB and UG-SLRFB respectively. The electrolyte was then pumped through the peristaltic pump inside the SLRFBs individually. The charge and discharge test has been carried in a constant current mode where a constant current of 500 mA. The charge/discharge data was collected in the voltage range of 0.9–2.2 V for rGO-SLRFB and 0.8–2.2 V for UG-SLRFB.

The rGO-SLRFB was assembled using rGO electrodes and the current collectors supported with acrylic endplates. The series of 20 charge/discharge cycles were performed and each cycle consisted of 24 min of charge followed by 15–20 min of

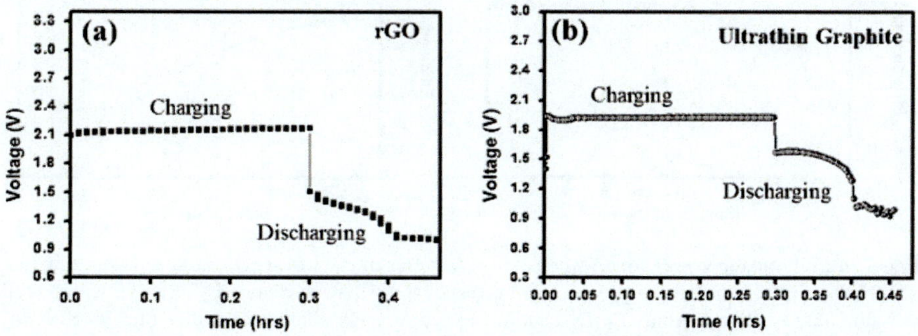

Fig. 4 **a** Charge/Discharge profile of rGO-SLRFB at 0.5A. **b** Charge/Discharge profile of the cell fabricated using UG-SLRFB at 0.5 A

discharge at 0.5 A of current. The voltage versus time response for 20 charge and discharge cycles is given (see Fig. 5a). The voltage of the cell was approximately 2.1 V during charge and 1.5 V during discharge. The discharge voltage is strongly affected by the mass transport of the reaction products away from the surface of the electrodes. Initially, in the formation cycles, the discharge capacity of rGO-SLRFB was low which increases and stabilized in the next few cycles (see Fig. 5c).

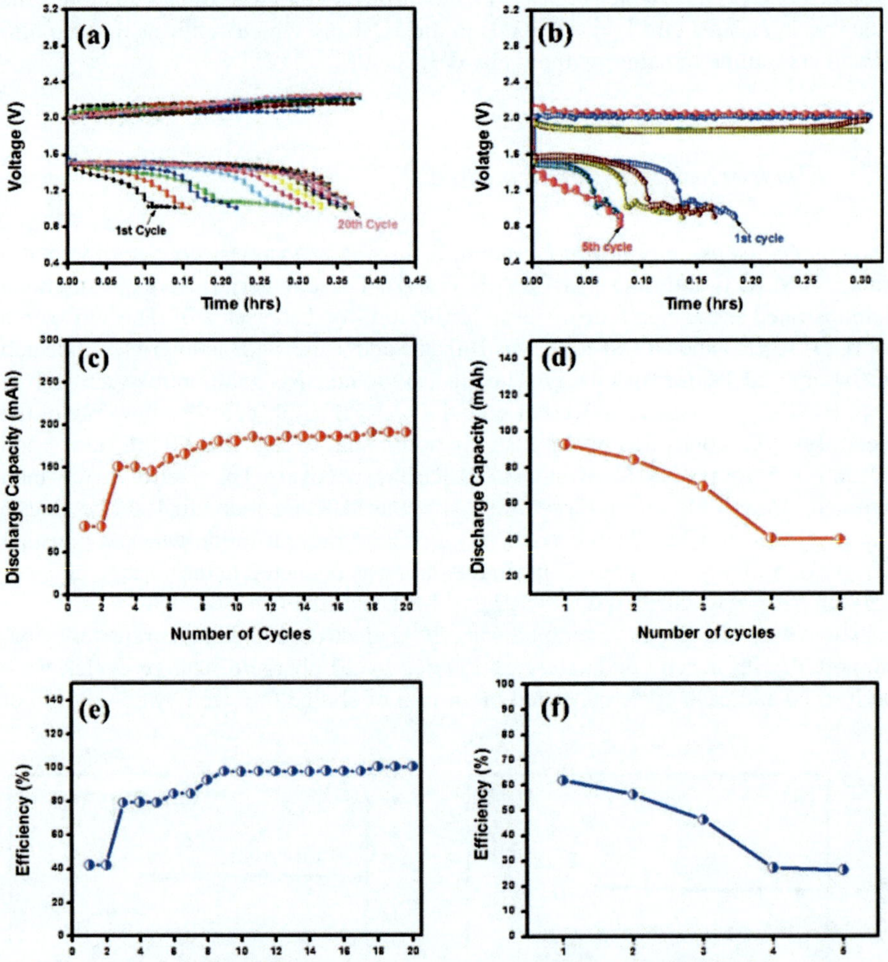

Fig. 5 **a** Cell voltage versus time response for 20 cycles of rGO-SLRFB single cell at a flow rate of 90 mL min^{-1}. **b** Cell voltage versus time response for five cycles of UG-SLRFB single cell at a flow rate of 20 mL min^{-1}. **c** Discharge Capacity versus number of cycles of rGO-SLRFB. **d** Discharge Capacity versus number of cycles of UG-SLRFB. **e** Efficiency versus number of cycles of rGO-SLRFB. **f** Efficiency versus number of cycles of UG-SLRFB

The charge/discharge profile of the first cycle is demonstrated (see Fig. 4a). The discharge plateau was observed at 1.5 V and continued until the end voltage of 0.9 V. After each subsequent cycle, the performance of the cell was found to improve. The discharge capacity and efficiency at the end of 20th cycle were calculated around 167 mAh and 92% (see Fig. 5c,e). The enhancement in the cell performance can be attributed to the high specific surface area and excellent conductivity of rGO material as compared to the materials studied earlier. The cell performance of rGO-SLRFB shows that it may be potential electrode materials for future high power SLRFB.

Similarly, UG-SLRFB was fabricated using UG electrodes which were also considered as the current collectors. The initial discharge capacity for UG-SLRFB was observed around 92.5 mAh and in later cycles, it degraded up to 40 mAh (see Fig. 5b,d), which can be attributed to the degradation of electrode materials on Ni-surface and deposition of lead oxide from the initial cycle and short circuit between the electrodes. The exposed edges Ni-substrate electrodes got corroded as it came in contact with acidic electrolytes, due to which the deposited ultrathin graphite layer partially removed from the Ni-surface and mixed in electrolyte solution due to which degradation in capacity failure has been observed in subsequent cycles.

Simultaneously the deposition of the lead and lead oxide began during the charging/discharging which is another reason for cell failure. At the end of the fifth cycle more deposition of lead and lead oxide on the surface of the electrodes was observed, moreover, the electrodes were short-circuited due to the overlapping of positive and negative electrodes due to deposition of lead on the electrode surface. Further, this resulted in fast voltage drop from the open-circuit voltage 1.5 V to the end voltage 0.8 V and therefore, resulting in lower average efficiency of around 20% (see Fig. 5f). This represented serious irreversibility which may be due to the short-circuiting and loss of electrode materials from the Ni-current collector. Due to this, the electrodes were damaged completely, which can be clearly seen from the optical images of lead deposited electrodes (see Fig. 6).

3.2 *Failure Analysis*

The electrochemical performance of UG-SLRFB was continued up to five cycles. The cut-down analysis of the cell was performed in order to understand and analyze the failure in cell which is due to the deposition over electrode surface and also to examine the change in the overall structure and behavior of the cell by opening the cell at the end of the 5th cycle and removing the deposits from the electrode surface, dried the deposits for further analysis of the deposited material (see Fig. 6). Due to the deposition, the inter-electrode gap gradually decreased and finally resulted in short-circuiting of the cell. This is the reason for the sudden drop of discharge capacity within five cycles. Therefore, to study the structural and morphological characteristics of the deposited material, XRD and SEM characterization was performed.

Fig. 6 **a** UG electrodes before testing of the cell. **b** UG electrodes after testing of the cell, both positive and negative electrodes overlapped each other

The positive side of the electrode was separated, as observed was completely damaged completely, especially the positive electrode. This is the reason the cell was not tested for more than five cycles. The lead deposition is polymorphic and consists of α-PbO_2 having orthorhombic structure and β-PbO_2 having tetragonal rutile structure. The identification of phase composition for α-PbO_2 and β-PbO_2 can be determined using X-ray diffraction. The XRD pattern for the electrode is shown in Fig. 7a. The phase identification was performed using ICDD database (α-PbO_2 card no. 72-2440 and β-PbO_2 card no. 76-0564) [24]. The diffraction peak is clearly distinguishable within the range of 20°–40°.

At the electrode, a rough powdery deposit was obtained due to overlapping of angular crystallites indicating lead oxide deposition favored by higher temperature range low charging current density [17]. The X-ray intensity peaks at ~25° [110], 32°

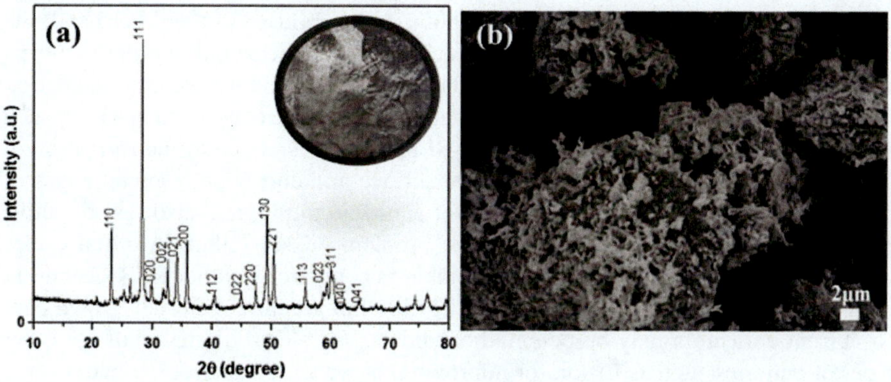

Fig. 7 **a** XRD of the deposited material. **b** SEM image of deposited material

[101], 36° [200], ~49.1° [211] describes the formation of lead oxide (see Fig. 7a). The appearance of the deposition is very compact and shiny.

To study the morphological behavior of the electrode deposited material SEM characterization was performed. The SEM images of dendrite formation are shown in Fig. 7b. The nano-crystalline structure of PbO_2 deposition has been reported previously in the range of 30–140 nm [25] and 30–50 nm [26]. At the electrode, random oval structure morphology was obtained and the maximum size of different oval structures was found to be around 40–60 μm. The deposited lead oxide was highly dense in nature which caused the formation of powder layer over the surface of the electrode. These types of large irregular morphology have been reported previously [25, 26].

The effect of deposition resulted in deteriorating battery life which reduced its efficiency to 20% at the end of 5th cycle. Sometimes the failure of the cell is not because of the lead deposit on the positive and negative electrode but may be due to lead deposited on non-conducting surface near to the active area of the electrodes [27]. Therefore, to eliminate that sludge kind of deposit, the different oxidizing or reducing agents may help in the reduction of the deposited material and also can improve the overall performance of the working electrodes.

4 Conclusion

In this study, we have designed, fabricated, and demonstrated the electrochemical performance of two different single-cell soluble lead redox system based on rGO coated electrodes and UG electrodes without using any membrane. The electrochemical performance of rGO-SLRFB has been studied and the efficiency was 92% at the end of 20th cycle, therefore, it can also be performed for higher number of cycles. As reported, the rGO material will make SLRFB much cheaper as compared to other RFBs in terms of design, fabrication and stability. On the other hand, the failure analysis of the UG-SLRFB single cell has been carried out in order to understand the characterization of the deposited material which in turn provides the reason for the early failure of the cell. The degradation of the electrochemical behavior such as capacity and efficiency in case of UG-SLRFB is caused due to the formation of lead dendrites/deposition on the electrode surface and corrosion of metal edges of the electrodes, which is the main cause for short-circuiting, and also loss of active material from the electrolyte. Solving the problem of cell failure in initial cycles will make this battery chemistry more suitable as compared to other RFB systems in terms of low cost and easy fabrication. The ultrathin graphite materials grown over Ni-current collectors may be potential candidates for electrode materials for other types of batteries such as Li-ion for improving the power density of the battery.

Acknowledgements This research work is funded and supported by Central Power Research Institute, under IHRD scheme of Ministry of Power.

References

1. Wu, X., Xu. H., Lu, L., Zhao, H., Fu. J., Shen, Y., Xu, P., Dong, Y.: PbO_2-modified graphite felt as the positive electrode for an all-vanadium redox flow battery. J. Power Sources **650**, 274–278 (2014)
2. Schrodt, J.P., Otting, W.T., Schoegler, J.O., Craig, D.: Trans. Electrochem. Soc. **90**, 281 (1946)
3. White, J., Powers, W., McMurtric, R., Pierce, R.: Trans. Electrochem. Soc. **91**, 21 (1947)
4. White, J., Balwin, J., Peebles, E., Powers, W.: Lead perchloric acid primary cell. United States Patent 2,492,206 (1949)
5. McDonald, G., Weissman, E., Roemer, T.: J. Electrochem. Soc. **119**, 660 (1972)
6. Beck, F.: Lead batteries. United States Patent 4,001,037 (1977)
7. Wurmb, R., Beck, F., Boehlke, K.: Secondary battery. United States Patent 4,092,463 (1978)
8. Henk, P.O., Piontkowski, Z.A.: Lead salt electric storage battery. US Patent 4,331,744 (1982)
9. Hazza, A., Pletcher, D., Wills, R.: A novel flow battery: a lead acid battery based on an electrolyte with soluble lead (II) Part I. Preliminary studies. Phys. Chem. Chem. Phys. **6**, 1773–1778 (2004)
10. Hazza, A., Pletcher, D., Wills, R.: A novel flow battery: a lead acid battery based on an electrolyte with soluble lead (II) Part IV. The influence of additives. J. Power Sources **149**, 103–111 (2005)
11. Pletcher, D., Wills., R.: A novel flow battery: a lead acid battery based on an electrolyte with soluble lead (II) Part II. Flow cell studies. Phys. Chem. Chem. Phys. **6**, 1779–1785 (2004)
12. Pletcher, D., Wills., R.: A novel flow battery: a lead acid battery based on an electrolyte with soluble lead (II) Part III. The influence of conditions on battery performance. J. Power Sources **149**, 96–102 (2005)
13. Pletcher, D., Zhou, H., Kear, G., Low, C.T.J., Walsh, F.C., Wills, R.G.A.: A novel flow battery—a lead-acid battery based on an electrolyte with soluble lead(II) Part VI. Studies of the lead dioxide positive electrode. J. Power Sources, **180**, 630–634 (2008)
14. Pletcher, D., Zhou, H., Kear, G., Low, C.T.J., Walsh, F.C., Wills, R.G.A.: A novel flow battery—a lead-acid battery based on an electrolyte with soluble lead(II) V. Studies of the lead negative electrode. J. Power Sources **180**, 621–629 (2008)
15. Collins, J., Kear, G., Pletcher, D.: A novel battery: a lead acid battery based on an electrolyte with soluble lead (II) Part VIII. The cycling of a 10 cm x 10 cm flow cell. J. Power Sources **195**, 1731–1738 (2010)

16. Banerjee, A., Saha, D., Guru Row, T.N., Shukla, A.K.: A soluble-lead redox flow battery with corrugated graphite sheet and reticulated vitreous carbon as positive and negative current collectors. Bull. Mater. Sci. **36**, 163–170 (2013)
17. Wallis, L.P.J., Wills, R.G.A.: Membrane divided soluble lead battery utilizing a bismuth electrolyte additive. J. Power Sources **247**, 799–806 (2005)
18. Lin, Y.T., Tan, H.L., Lee, C.Y.: Stabilizing the electrodeposit-electrolyte interphase in soluble lead flow batteries with ethanoate additive. Electrochim. Acta **263**, 60–67 (2018)
19. Daniel, C., Dmitry, V., Kosynkin, M., Jacob, M.: Improved Synthesis of Graphene Oxide. ACS Nano. **4**, 4806–4814 (2010)
20. Kuldeep, R., Kim, S.D., Ahn, J.-H.: Additive-free thick graphene film as an anode material for flexible lithium-ion batteries. Nanoscale, **7**, 7065–7071 (2015)
21. Kumar, C.M.P., Venkatesha, T.V., Shabadi, R.: Preparation and corrosion behavior of Ni and Ni–graphene composite coatings. Mater. Res. Bull. **48**, 1477–1483 (2013)
22. Li, Y., Tang, L., Li, J.: Preparation and electrochemical performance for methanol oxidation of pt/graphene nanocomposites. Electrochem. Commun. **11**, 846–849 (2009)
23. Kuldeep, R., NagaMahesh, M., Singh, V., Ahn, J.H.: Synthesis of additive free electrode material of supercapacitor for energy storage applications. In: International Conference on Industrial and Information Systems (ICIIS), pp. 448–451 (2016)
24. Li, X., Pletcher, D., Walsha, F.C.: Electrodeposited lead dioxide coatings. Chem. Soc. Rev. **40**, 3879–3894 (2011)
25. Sires, I., Low, C.T.J., Ponce-de-Leon, C., Walsh, F.C.: The characterstics of PbO_2-coated electrodes prepared from aqueous methanesulfonic acid under controlled deposition conditions. Electrochim. Acta, **55**, 2163–2172 (2010)
26. Mehdinia, A., Mousavi, M.F., Shamsipur, M.: Nano-structured lead dioxide as a novel stationary phase for solid-phase microextraction. J. Chromatogr. **1134**, 24–31 (2006)
27. Walsh, F.C.: A first course in Electrochemical Engineering. Electrochemical Consultancy. ISBN: 0.9517307 11 (1993)

Development of Novel Ternary SnSb-ZnO Nanocomposites as Alternative Anode Material for Lithium-Ion Batteries

Shivangi Rajput and Amrish K. Panwar

Abstract Novel Ternary Sn-Sb-ZnO nanocomposites were synthesized as alternative anode materials for Lithium-ion batteries (LIBs) to replace well known graphitic carbon anode materials. In this study, SnSb-(ZnO)x active phase alloy composites were synthesized via physical route where x corresponds to different weight percentage of ZnO (x = 0%, 1%, 2.5%, 15%). Physicochemical characterization of the synthesized materials has been performed using XRD, SEM, and EDS to confirm proper phase formation, particle size, and determination of elemental composition, respectively. The electrochemical properties were analyzed by cyclic voltammetry (CV), electrochemical impedance spectroscopy (EIS). Further, a simulated model has also been developed by taking inputs from the 2D lithium-ion battery model available in COMSOL multi-physics.

Keywords Anode material · SnSb-ZnO · Lithium-ion batteries · Cyclic voltammetry

1 Introduction

In the last few decades, there has been tremendous growth in the field of batteries. LIBs have been proved to be a promising one in many portable devices such as mobile phones, laptops, and microelectronic devices. LIBs have gained so much attention due to its obvious advantages like superior energy densities, long life cyclability, low memory effect, and low environmental effects. In general, the most important component of a battery which decides the transport of charges and storage of energy is electrode material. The electrode material is key to achieve high energy or power density, high capacity, long cyclic stability, and high coulombic efficiency [1]. Since the first commercialization of LIBs in the year 1990–91, Graphite has been widely used in commercial LIBs, but it also has certain drawbacks like low gravimetric

S. Rajput · A. K. Panwar (✉)
Department of Applied Physics, Delhi Technological University, Delhi, India
e-mail: amrish.phy@dtu.ac.in

Y. Sharma et al. (eds.), *Recent Research Trends in Energy Storage Devices*,
https://doi.org/10.1007/978-981-15-6394-2_17

capacity and volumetric capacity and safety concerns which makes graphite unsuitable for next generation of LIBs. Moreover, carbonaceous electrode materials due to its low theoretical capacity cannot alone meet the increasing demand for modern power requirements [2]. Therefore, there is an utmost need for alternative novel anode materials for high lithium storage capability and safe operation for next-generation LIBs. The group IV elements such as Silicon, Germanium, tin have been considered to be promising anode materials for LIBs due to their higher specific capacities as compared to carbon-based graphitic anode material [3–12]. Among the above-mentioned anode candidate, tin (Sn) has received much attention because of its high theoretical specific capacity, 994 $mAhg^{-1}$ [13–16]. Alloy anode materials, especially: anode materials based on or form compounds with Sn- and Sb- have presented themselves as a possible and sustainable replacement of graphite. Lithium forms an alloy with elements Sn and Sb in the form of $Li_{4.4}Sn$ and Li_3Sb and resulting in their theoretical specific capacities nearly equal to 991 mAh/g and 660 mAh/g, respectively. This makes a wide range of tin-based compounds to be considered as potential anode material. Antimony being a close neighbor of tin is also a potential anode material in the rudimentary form as well as when it forms compounds [17].

Composite alloy anodes such as SnSb is an active-matrix composite in which both the component as well as the matrix are active (i.e., all components are reactive towards lithium-ion during different stages in charge/discharge cycles). In such a composite material, while one component undergoes lithiation, the other acts as a buffer to relieve the volume change, as each has a different onset potential for reaction with Lithium.

In SnSb alloy, Sb starts lithium-insertion at 0.9 V versus Li/Li^+ while Sn contains the volume expansion of the lithiated product. Here, Sn has an onset potential of 0.6 V [18, 19]. Superior cycling stability of the active composite anode can be achieved by restricting the charge to a restricted state of lithiation of the second component. As a result, the initial irreversible capacity is also improved. The cyclability of nano-SnSb is ascertained to be better than individual Sn and Sb anodes. With a high theoretical capacity of 978 $mAhg^{-1}$, Zinc Oxide (ZnO), among various transition metal oxides, is considered to be a potential anode material for Lithium-ion batteries. Its poor cyclability due to volume expansion (228%) is compensated by the use of nanostructured materials that accommodate large strain as well as provide shorter diffusion lengths for Li-ions [20].

In this study, synthesis and physicochemical characterization of SnSb-ZnO nanocomposite have been reported. The synthesis of SnSb-$(ZnO)_x$ [$x = 0, 1, 2.5, 15$ wt%] was performed via high energy ball milling technique. Physicochemical characterization of the synthesized materials has been performed using XRD, SEM, and EDS. The cyclic voltammetry (CV) and electrochemical impedance spectroscopy (EIS) was carried out to study the electrochemical performance of bare and optimized SnSb-ZnO. Further, a simulated model has also been developed by taking inputs from the 2D lithium-ion battery model available in COMSOL multi-physics.

2 Experimental

2.1 Synthesis

All precursors used in the synthesis were of analytical grade. Nanocomposites of SnSb-ZnO were synthesized by the mechanochemical activation technique. High energy ball milling was used to synthesize nano-sized SnSb-$(ZnO)_x$ ($x = 0, 1, 2.5, 15$ wt%) composite material. The starting materials, Sb_2O_3 (99% wt) and Sn powders (99.9% wt) are mixed in the molar ratio 1:1 and further, the different weight % of ZnO were mixed and subjected to ball milling for 18 h. The reduced material was then obtained and subjected to thermal treatment in the vacuum oven for 12 h at 160 °C. The product was cooled and the desired SnSb-ZnO nanocomposite was obtained.

2.2 Physical Characterization

The phases of the as-synthesized composite material were analyzed using PANanlytical make X'pert PRO Diffractometer with Cu-$K_{\alpha 1}$ radiations of wavelength, $\lambda =$ 1.540 Å in wide range scan of 10°–80°. The morphology and the microstructures of the particles were examined using ZEISS to make SUPRA 55 model of FE-SEM. Thermogravimetric analysis (TGA) was done from room temperature to 500 °C with a heating rate of 10 °C min^{-1}.

2.3 Electrochemical Characterization

To perform the electrochemical measurements, the electrodes were prepared by mixing 80 wt% active material (SnSb-ZnO), 15 wt% polyvinylidene fluoride as binder, and 5 wt% acetylene black (99.5% purity) as a conducting agent in 1-methyl-2-pyrrolidinone (NMP) to form a slurry. The resultant slurry was pressed onto a copper foil and dried in a vacuum oven for 12 h at 120 °C. The resulting film was used to prepare circular electrodes with active material of SnSb-ZnO with the loaded material of approximately 1.2 g/cm^2. The coin-type cells were assembled in a glove box (MBRAUN, UNILab 2000), using lithium metal foil as the counter electrode. CELGARD-2400 used as the separator and 1 M $LiPF_6$ were dissolved in a solvent mixture of EC DMC and DEC (1:1:1 v/v) as an electrolyte. All the electrochemical measurements were performed at room temperature.

3 Result and Discussion

3.1 Structural and Morphological Analysis

X-ray diffraction (XRD) analysis was carried out to determine the structure and crystal phase of the as-synthesized sample. Figure 1 shows the characteristics peaks of the SnSb-(ZnO)-$_x$ (x = 0, 1.0, 2.5 and 15 wt%) nanocomposite. The peaks indexed for SnSb-ZnO indicate β-SnSb phase corresponds to the rhombohedral structure with the space group, $R\bar{3}m$. The sharp peaks which are observed in the diffraction pattern are those of SnSb, ZnO, and Sn along with a few weak peaks for SnO_x, SbO_x, and SbZn. With the increase in the ZnO content in the composite, the peaks for SnSb and Sn are steadily diminished and that of ZnO is intensified. In Fig. 2a–d can be seen in the SEM images of SnSb and SnSb-ZnO nanocomposite. The results indicate very small particles of SnSb and SnSb-ZnO.

The obtained images show homogenously distributed tiny irregular particles which also seem to have undergone uniform aggregation. The elemental analysis of each sample, in order to verify each component present in the sample along with their relative abundance was detected using EDS microanalysis as shown in Fig. 2a–d.

From the EDS microanalysis results, the presence of each elemental component, i.e., Sn, Sb in bare SnSb sample and Sn, Sb, Zn, and O in SnSb-ZnO samples were estimated. In the quantitative analysis of each sample, the weight percentage and atomic percentage of individual components were revealed. Bare SnSb has 43.35% Sn, 46.66% Sb, and 7.11% O content. Similarly, SnSb-$ZnO_{x=1\ wt\%}$ has 33.92% Sn, 36.01% Sb, 0.69% Zn, and 29.38% O content. SnSb-$ZnO_{x=2.5\ wt\%}$ has 35.68% Sn, 33.12% Sb, 1.01% Zn, and 30.18% O content. SnSb-$ZnO_{x=15\ wt\%}$ was observed to contain 39.20% Sn, 32.54% Sb, 2.10% Zn, and 26.16% O content.

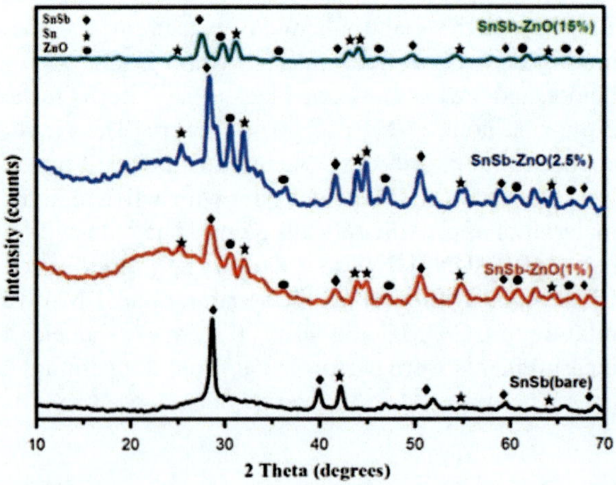

Fig. 1 Pattern obtained by X-ray diffraction analysis of bare SnSb and SnSb-ZnO composite

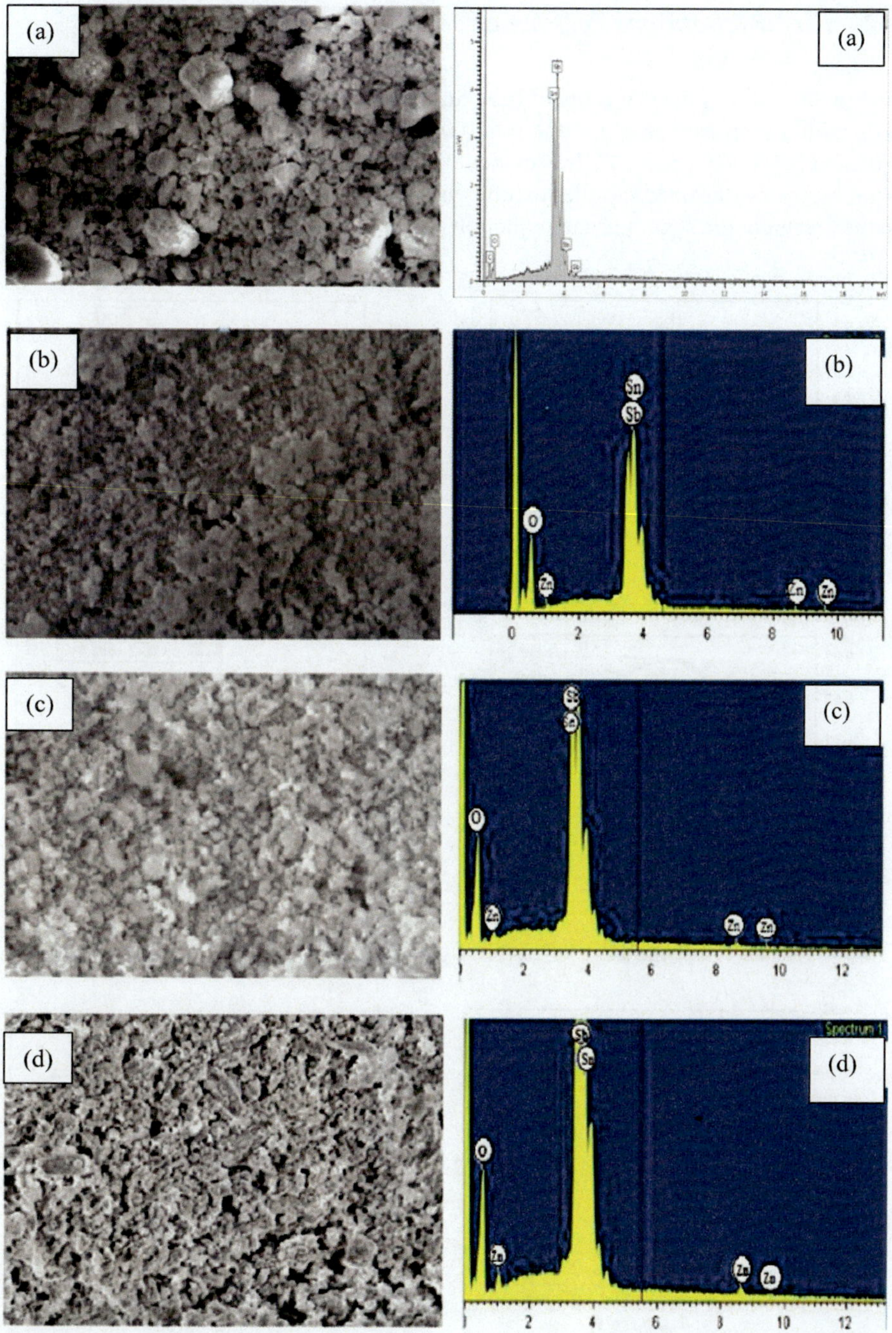

Fig. 2 Images of SEM and EDX analysis of **a** bare SnSb, **b** SnSb-ZnO (wt 1%), **c** SnSb-ZnO (wt 2.5%) and **d** SnSb-ZnO (wt 15%)

3.2 Electrochemical Performance

Figure 3a–c shows the CV plots of bare SnSb and SnSb-ZnO nanocomposite materials with a scan carried out at the rate 0.05 mV s^{-1} over a voltage scope between 0.01 and 3.5 V versus Li/Li^{+}. It is evident from the plot that over the given potential window, the synthesized samples exhibit only the characteristic redox peaks of SnSb. In the first cathodic scan, a broad reduction peak can be observed at around 1.3–1.5 V,

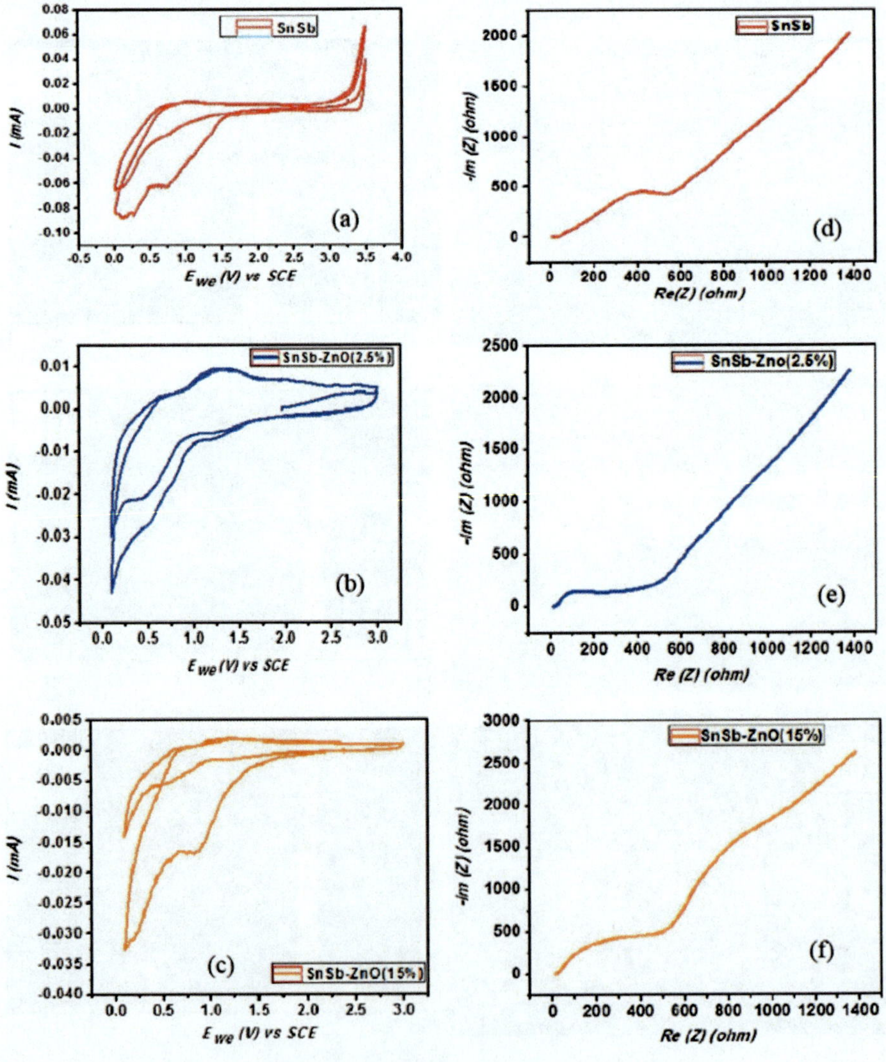

Fig. 3 CV curves of **a** bare SnSb, **b** SnSb-ZnO (wt 2.5%) and **c** SnSb-ZnO (wt 15%) and EIS curves of **d** bare SnSb, **e** SnSb-ZnO (wt 2.5%) and **f** SnSb-ZnO (wt 15%)

which is seen to disappear in subsequent cycles. The peak could be attributed to the disintegration of the electrolyte and the formation of the layer of SEI (solid electrolyte interface). A significant discharge (reduction) peak is observed at approximately around 0.75 V which is persistent in the following cycles. This peak can be attributed to the lithiation reaction resulting in the formation of Li_3Sb and the release of metallic tin. The broad reduction peak around 0.35–0.4 V corresponds to the alloying reaction between Sn and Li, resulting in the formation of Li_xSn. The peaks appearing after 0.66 V confirm the reaction of ZnO with Li which is very obscure. During the anodic scan, the de-lithiation of Li_xSn during the first cycle can be ascribed to the oxidation peaks close to 0.6 V. Similarly the presence of an oxidation peak around 1.1 V is accredited in the de-alloying process leading to the conversion of Li_3Sb to Sb and resulting in the formation of SnSb alloy. Good electrochemical reversibility is implied by the presence of stable and reversible peaks in the second and third cycles. The difference in the CV plot for SnSb and SnSb-ZnO composite results from the involvement of ZnO in the intercalation reaction. The overall reactions involving lithiation and de-lithiation of bare SnSb and SnSb-ZnO composite are as follows:

$$\begin{aligned} &3\mathrm{Li} + \mathrm{SnSb} \rightleftharpoons \mathrm{Li_3Sb} + \mathrm{Sn} \\ &x\mathrm{Li} + \mathrm{Sn} \rightleftharpoons \mathrm{Li}_x\mathrm{Sn} \quad (x \le 4.4) \\ &2\mathrm{Li} + \mathrm{ZnO} \rightleftharpoons \mathrm{Zn} + \mathrm{Li_2O} \\ &\mathrm{Zn} + x\mathrm{Li} \rightleftharpoons \mathrm{Li}_x\mathrm{Zn} \end{aligned}$$

In order to better explain the electrochemical behavior of the bare SnSb and SnSb-ZnO, the ac impedance plots were recorded before cycling. EIS was conducted in the scanning frequency range of 10 MHz–10 kHz at an ac voltage amplitude of 5 mV.

The impedance spectra are observed to form a dampened arc or semi-circle in the region spanning from high to medium frequency and a sloping linear graph in the region of low frequency of "Nyquist plot". The Nyquist plots thus obtained are fitted by the equivalent circuit in the insets of Fig. 3d–f. The intercept of the semi-circular arc on the Z' axis present in high-frequency region corresponds to the electrolyte resistance and is denoted by Rs in the equivalent circuit. The semi-circle prominent in middle-frequency section represents charge transfer resistance (Rct) corresponding to several reactions occurring at the electrode/electrolyte interface. CPE represents the constant phase angle elements which are associated with the double-layer capacitance. We refer to the Warburg resistance, and it represented by the inclined line present across the low-frequency zone. It is related to the Li-ion bulk diffusion resistance. The increase in the ZnO percentage leads to an increase in the radius of the semi-circle. However, the intercept observed in the region of high frequency and the linear component does not vary much with the ZnO content. The results of fitting of R_s and R_{ct} are summed up in Table 1.

Table 1 Impedance parameters of SnSb-ZnO nanocomposites at open-circuit voltage

$SnSb\text{-}ZnO_x$	x = 0%	2.5 wt%	15 wt%
R_s	73.48	6.507	33.57
R_{ct}	748.9	242.2	770.5

4 Simulation Results

The simulation shows the distribution of the concentration of lithium on the surface of the electrode particles by the end of 2700 s. The initial lithium concentration in the porous electrode is 1000 mol/m^3. Throughout the discharge process, as lithium is dissolved on the negative lithium metal electrode, the lithium concentration in the porous electrode increases. Electrolyte salt concentration in the simulated model is depicted in Fig. 4, whereas the concentration of lithium is shown in Fig. 5. The concentration in the positive electrode particles at two selected locations within the porous electrode is displayed in Fig. 6. The green line displays the concentration at the location furthest away from the current collector and shows that less discharge has taken place there (concentration lower). Figure 7 shows the time-dependent cell voltage predicted by the model during the discharge simulation.

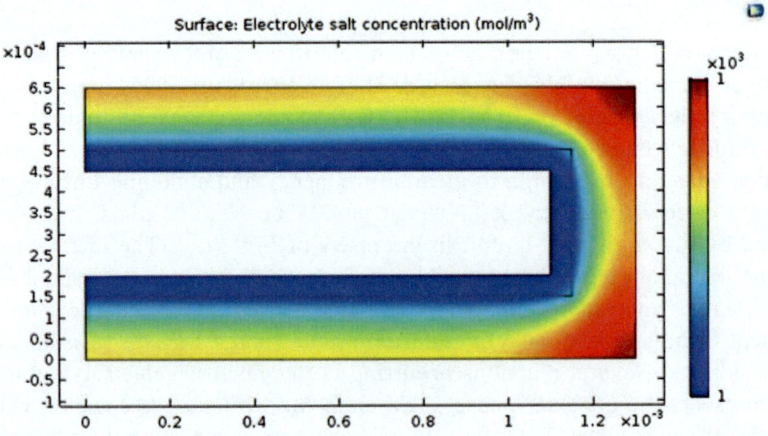

Fig. 4 Concentration of electrolyte at the surface of electrode particles

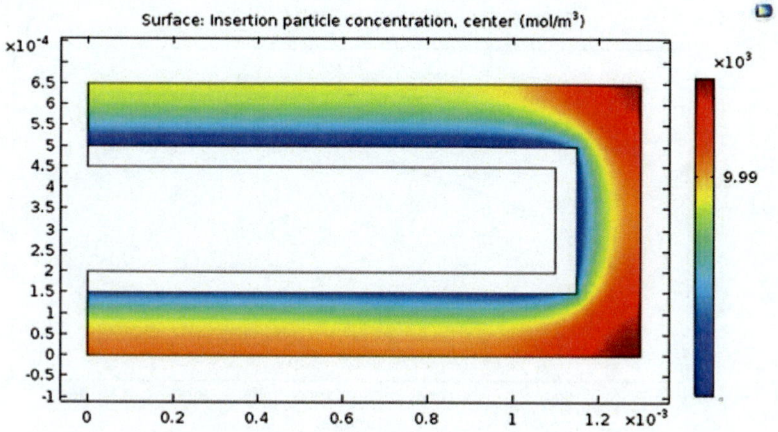

Fig. 5 Concentration of lithium at the surface of electrode particles

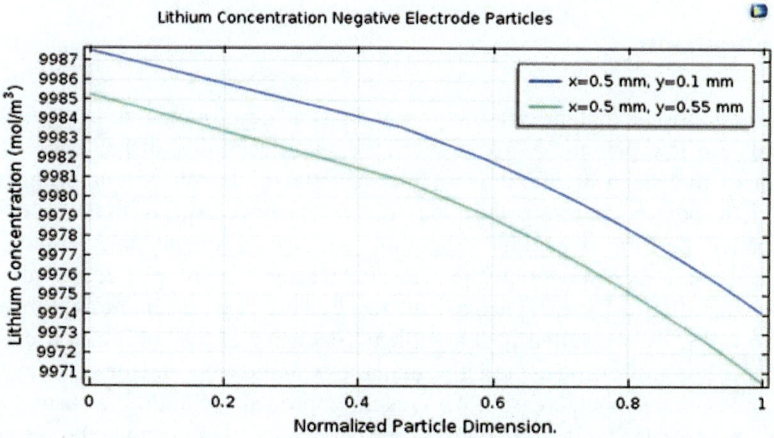

Fig. 6 Lithium concentration in the electrode particles at two selected positions, within the model geometry

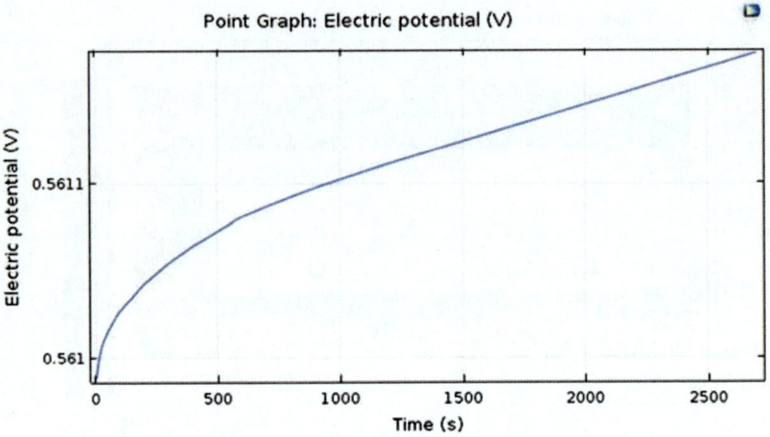

Fig. 7 The time-dependent cell voltage predicted by the model during the discharge simulation

5 Conclusion

It is well established that the criteria for selection and development of practical anode materials are low potential, specific capacity, and cyclability. It has been focused throughout that none of the above-mentioned standards can be compromised on behalf of the others. Therefore, the aim is to develop anode material that would lead to better battery performance with low potential window. In conclusion, a ternary SnSb-ZnO_x active-matrix nanocomposite was synthesized via the physical route (high energy ball milling). The presence of ZnO resulted in a decrease in the agglomeration of SnSb alloy. This resulted in a significant decrease in the lithium-ion diffusion distance, lithium-insertion in the active materials was reduced, and the space available for volume expansions increased. As a result, improved cyclability is observed. The CV curves verify the multistep lithiation and de-lithiation occurring in the composite. The electrochemical performance establishes the nanocomposite as an important anode material owing to the formation of stable "solid electrolyte interface film" as well as "excellent charge/discharge profiles".

References

1. Liu, L., Xie, F.: Tin-based anodes materials with well-designed architectures for next-generation lithium ion batteries. J. Power Sources, **321** (2016)
2. Tian, H., Xin, F.: High capacity group-IV elements (Si,Ge,Sn) based anodes for Li ion batteries. J. Materiomics **1** (2015)
3. Armand, M., Tarascon, J.M.: Building better batteries. Nature (2008)
4. Goriparti, S., Miele, E., De Angelis, F., Di Fabrizio, E., Zaccaria, R.P., Capiglia, C.: Review on recent progress of nanostructured anode materials for Li-ion batteries. J. Power Sources (2014)

5. Terranova, M.L., Orlanducci, S., Tamburri, E., Guglielmotti, V., Rossi, M.: Si/C hybrid nanostructures for Li-ion anodes: an overview. J., Power Sources (2014)
6. Wang, X.H., Susantyoko, R.A., Fan, Y., Sun, L.M., Xiao, Q.Z., Zhang, Q.: Vertically aligned CNT-supported thick Ge films as high-performance 3D anodes for lithium ion batteries (2014)
7. Sourice, J., Quinsac, A., Leconte, Y., Sublemontier, O., Porcher, W., Haon, C., et al.: One-step synthesis of Si@C nanoparticles by laser pyrolysis: high-capacity anode material for lithium-ion batteries. ACS Appl. Mater. Inter. (2015)
8. Ryu, J., Choi, S., Bok, T., Park, S.: Nanotubular structured Si-based multicomponent anodes for high-performance lithium-ion batteries with controllable pore size via coaxial electro-spinning. Nanoscale (2015)
9. Yu, B.C., Hwa, Y., Park, C.M., Kim, J.H., Sohn, H.J.: Effect of oxide layer thickness to nano-Si anode for Li-ion batteries. Rsc Adv. (2013)
10. Jeong, S, Lee, J.P., Ko, M., Kim, G., Park, S., Cho, J.: Etched graphite with Internally Grown Si nanowires from pores as an anode for high density Li-ion batteries. Nano Lett. (2013)
11. Nguyen, H.T., Yao, F., Zamfir, M.R., Biswas, C., So, K.P., Lee, Y.H., et al.: Highly interconnected Si nanowires for improved stability Li-ion battery anodes. Adv. Energy Mater. (2011)
12. Sethuraman, V.A., Chon, M.J., Shimshak, M., Van Winkle, N., Guduru, P.R.: In situ measurement of biaxial modulus of Si anode for Li-ion batteries. Electrochem. Comm. (2010)
13. Wang, B., Luo, B., Li, X., Zhi, L.: The dimensionality of Sn anodes in Li-ion batteries. Mater. Today **15**(12) (2012)
14. Idota, Y., et al.: Tin-based amorphous oxide: a high-capacity lithium-ion-storage. Mater. Sci. **276**, 1395 (1997)
15. Ibarz, A., et al.: Electronic structure of host lattices for intercalation compounds: SnS_2, $SnSe_2$, ZrS_2, and TaS_2.Chem. Mater. **10**, 3442 (1998)
16. Seo, J.W., et al.: Two-dimensional SnS_2 nanoplates with extraordinary high discharge capacity for lithium ion batteries. Adv. Mater. **20**, 4269 (2008)
17. Goriparti, S., Miele, E., Angelis, F., Fabrizio, E.: Review on recent progress of nanostructured anode materials for Li-ion Batteries. J. Power Sources **257** (2014)
18. Yang, J., Winter, M., Besenhard, J.O.: Small particle size multiphase Li-alloy anodes for lithium-ionbatteries. Solid State Ionics **90**(1–4), 281–287 (1996)
19. Szczech, J.R., Jin, S.: Nanostructured silicon for high capacity lithium battery anodes. Energy Environ. Sci. **4**(1), 56–72 (2011)
20. Li, Y., Zhang, W., Cai, H., Wang, J.: SnSb-ZnO composite materials as high performance anodes for Li-ion batteries. RSC Adv. **5**, 105643 (2015)

Structural, Electrochemical, Thermal and Electrical Performance of $LiMnPO_4$ Cathode Material for Lithium-Ion Batteries

Snigdha Sharma, Amrish K. Panwar, and M. M. Tripathi

Abstract In this investigation, olivine type pristine $LiMnPO_4$(LMP) and Magnesium (Mg) substituted $LiMn_{1-x}Mg_xPO_4$ ($x = 0$ to 1) solid solution was synthesized using the solid-state route. Thermogravimetric analysis (TGA) of pristine LMP indicates the transition temperature to be around 320 °C which correspond to the release of organic compounds during the evaporation process and another transition at 450 °C corresponds calcination for proper phase. Further, structural characterization by XRD is carried out to confirm the proper phase formation of pure LMP and Mg-doped LMP. SEM and EDX were also attempted to visualize the morphological distribution, shape, size of the particles and stoichiometric composition at the final phase, respectively. Moreover, electrical conductivity of 3% Mg-doped LMP sample was carried out to study the activation energy of the synthesized sample. The electrochemical performances such as CV, EIS and GCD of 3% Mg-doped LMP were also measured along with the simulation results to estimate the behaviour of cathode during charge and discharge theoretically.

Keywords Cathode material · Lithium-ion batteries · $LiMnPO_4$ · Cyclic voltammetry

1 Introduction

The progress in outstanding-performance of Lithium-Ion Batteries (LIBs) with high energy density, lightweight, high cell voltage and long cycle life is vital because of the increasing surplus demand of portable electronics and electric vehicles (EVs) [1, 2]. Among various types of materials used as cathode reported for lithium-ion batteries, olivine type LMP [3] is one of the most promising materials because of its high potential [3] and outstanding experimental specific discharge capacity (>200 mAh/g) [3]

S. Sharma · M. M. Tripathi
Department of Electrical Engineering, Delhi Technological University, Bawana Road, Delhi, India

S. Sharma · A. K. Panwar (✉)
Department of Applied Physics, Delhi Technological University, Bawana Road, Delhi, India
e-mail: amrish.phy@dtu.ac.in

Y. Sharma et al. (eds.), *Recent Research Trends in Energy Storage Devices*,
https://doi.org/10.1007/978-981-15-6394-2_18

comparatively to $LiCoO_2$ cathode [4] available commercially. The most promising batteries which have received great attention nowadays are lithium transition metal phosphates [2, 5, 6] i.e. $LiMPO_4$ (M=Fe, Mn, Co, Ni). These batteries are having better thermal stability of electrochemical performance, low toxicity, and low cost [7]. Despite high potential and high electrochemical performance low intrinsic/ionic electronic conductivity [5, 8] is the main drawback of LMP and non-oxidizing nature of Mn ions [9]. However, several efforts have been made worldwide to overcome all these shortcomings in recent years. On improving, intrinsic/ionic electronic conductivity is minimizing the size of particles (nanomaterial) and uniform particle size distribution with preferable morphology. Therefore, the substitution of multi transition metal $LiMn_xMg_{(1-x)}$ PO_4 (M=Fe, Mg, V) [10, 11], has been attempted to improve the reaction kinetics [12, 13] of $LiMnPO_4$ (LMP) without the energy density loss [11, 12, 14].

In this work, we have focused on the substitution of Mg in LMP to facilitate the crystal structure as well as to enhance the electrochemical performances. The pristine LMP and Mg-doped LMP at different wt% were synthesized and characterized to perform the XRD, SEM and EDX to check the proper phase formation. Electrical conductivity and electrochemical characterizations were also performed for the optimized Mg-doped LMP to study activation energy, electrochemical performance and kinetics for lithium-ion batteries.

2 Experiment

2.1 Synthesis of ($LiMnPO_4$)

Synthesis of $LiMnPO_4$ was carried out by the stoichiometric amount of compounds including lithium hydroxide [$LiOH{\cdot}H_2O$], Manganese Oxide [MnO], di-Ammonium hydrogen orthophosphate [$(NH_4)_2HPO_4$] using the conventional solid-state route. All the above precursors were mixed for 4 h and grounded in mortar-pestle. Calcination was performed at 450 °C for 2 h in a tubular furnace at an inert atmosphere to form the homogenous solid solution. This led to the removal of volatile impurities in gaseous form. Thus, the sample obtained was abbreviated as LMP. Furthermore, $LiMnPO_4$ (LMP) is doped with different weight % of Mg (1, 3%), as $LiMn_{0.99}Mg_{0.01}PO_4$ (LMMP1) and Li $Mn_{0.97}Mg_{0.03}PO_4$ (LMMP3), respectively.

2.2 Materials Characterization of Synthesized Samples

Thermogravimetric analysis of synthesized pure LMP and Mg-doped LMP was measured by SII 6300 EXSTAR model [15] of TGA in an inert atmosphere with a 5 °C min^{-1} ramping rate. Phase confirmation of the prepared composition was observed from X-ray diffractometer model: D8 Advance made by Bruker, Germany. X-ray radiation of $CuK\alpha_1$ is used to observe XRD patterns. Microstructural analysis of the samples was done by using Hitachi S-3700 SEM [16] (scanning electron microscope). Electrical characterization of the synthesized sample was observed by using KEITHLEY 6430 SUB-FEM TO AMP REMOTE Source Meter in the voltage window 1–2 V and temperature range 30–300 °C with a 2 °C/min heating rate. DC conductivity (σ_{DC}) is calculated with the help of resistivity. The resistivity can be calculated with the help of given equations:

$$\rho = R * \left(\frac{R}{L}\right) \quad (1)$$

where

- R Resistance of the sample (Ω),
- ρ Resistivity of the material (Ω^{-1}),
- L Thickness of the pellet (cm),
- A Cross-section Area (cm^2).

As we get the value of resistivity, then DC conductivity (σ_{DC}) is calculated by reciprocating the resistivity of the material in the given equation:

$$\sigma_{DC} = 1/\rho \quad (2)$$

Activation Energy (E_a) can be calculated with the help of the Arrhenius equation, by placing the values of conductivity and temperature in the below-given equation:

$$\sigma = \sigma_o \exp(-E_a/K_B T) \quad (3)$$

where

- E_a is activation energy (eV),
- σ is electronic conductivity (S/cm),
- T is temperature (K),
- K_B is Boltzmann constant ($J \cdot K^{-1}$).

2.3 *Preparation of Electrode*

To perform electrochemical measurements, CR2032 coin half cells were assembled in MBRON Germany make glove box work station under inert atmosphere ≤ 0.5 ppm H_2O/O_2 level. Electrodes LMP, LMMP1 and LMMP3 were prepared by 75% of its weight of active material. 5% of super P and 20% of polyvinylidene difluoride (PVDF) were added as a binder for slurry formation. The synthesized mixture was continuously stirred for 12 h at room temperature and then consistently coated on Al-foil current collector with the help of "doctor blade" technique [17]. This coated aluminium foil was dried in a vacuum oven at 120 °C and kept for overnight so that the solvent gets evaporated and then pressed with a rolling press machine at 50 mPa. The requisite mass/capacity balancing condition was achieved by the desired mass loading obtained by varying the coating thickness. The active loading of the mass obtained was 3.4 ± 0.2 mg while the thickness of the cathode electrodes was 51 ± 3 μm which provides original loading of the mass to be around 2.5 ± 0.2 mg. The electrode formed was heated under vacuum condition at 80 °C to remove the entire moisture from the material. Assembling of CR2032 coin cells is required for the electrochemical performance of the synthesized samples inside the glove box. Half cells with cathode materials LMP, LMMP1 and LMMP3 were fabricated. The reference and counter electrode were formed utilizing metallic lithium and separator was inserted utilizing microporous polypropylene sheet Celgard 2400 [18]. 1 M $LiPF_6$ in EC: DMC 1:1 v/v was utilized as an electrolyte [19].

2.4 Electrochemical Characterization

Cyclic voltammetry (CV) measurements were performed at a scan rate of 0.05 mV s^{-1} within the probable range of 2.0–4.0 V (vs. Li/Li^+ metal) and Electrochemical Impedance Spectroscopy (EIS) were performed at room temperature (298 K) within the frequency range of 10 mHz–100 kHz. The galvanostatic charge/discharge (GCD) tests were performed to estimate the capacity of 3% Mg-doped optimized LMP for all the cells assembled. 170 $mAhg^{-1}$ was assumed as the theoretical capacity for LMP. Biologic Science Instruments make VMP3 model [20] at 298 K were utilized to perform every electrochemical test.

3 Results and Discussions

3.1 Thermal Analysis

The results of the thermogravimetric analysis of the ball-milled $LiMnPO_4$ compound show the mass loss stages during the heating process (Fig. 1).

The initial stage of transition temperature was around 320 °C which is associated with the release of organic compounds during evaporation processes. The other stage of mass loss around 450 °C is associated with a calcination temperature resulting in the final olivine type phase of $LiMnPO_4$ (LMP).

Fig. 1 Thermogravimetric analysis curve

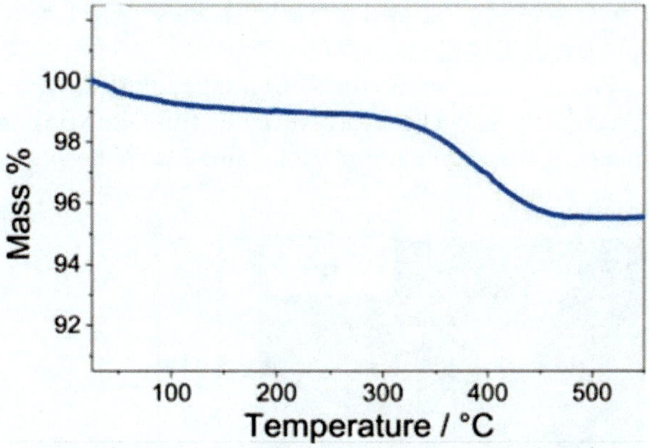

3.2 Phase and Morphological Analysis

The XRD patterns of the as-synthesized samples of LMP, LMMP1 and LMMP3 are shown in Fig. 2. All the diffraction peaks for olivine- LMP, LMMP1 and LMMP3 samples belong to orthorhombic phase formation with space group pnma. The sharp peaks of all the synthesized samples were crystalline. Major peaks of XRD patterns of all the samples LMP, LMMP1 and LMMP3 was observed at $2\theta = 16.92°$, 22.45°, 25.31°, 29.24°, 35.193°, 52.42° and 61.121° and indicate no shifting leading to the proper phase formation after the doping of Mg as 1 and 3 wt%. However, the peak intensity bit reduces at the sample LMMP3 and peak broadening also which reveals the crystallite size may increase a bit as compared to the pristine and Mg-doped

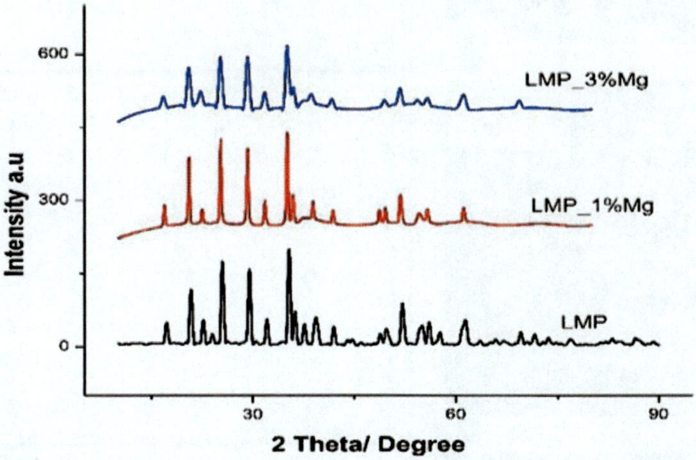

Fig. 2 XRD patterns of pristine LMP, LMMP1 and LMMP3 samples treated at 450 °C for 2 h in inert atmosphere

1 wt%. It is found all the peaks in these XRD patterns match with the standard database ICDD file no. 00-033-0804.

Figure 3a, c, e shows the SEM images of the LMP, LMMP1 and LMMP3 samples, respectively. It can be clearly revealed from the SEM images that all the samples have spherical shape type morphology along with the slight agglomeration in the particle

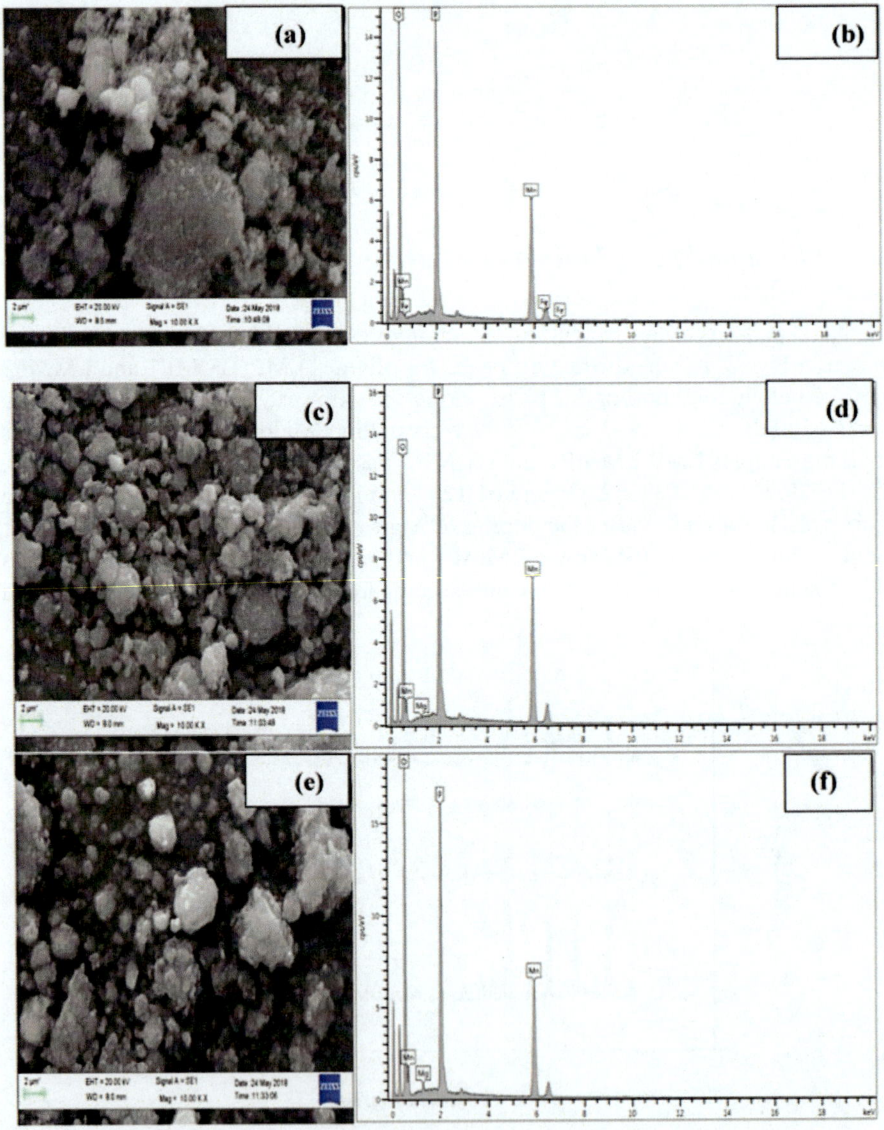

Fig. 3 SEMs and EDS images **a**, **b** pristine LMP, **c**, **d** LMMP1 and **e**, **f** LMMP3, respectively

and non-uniform distribution. The average particle size of LMP and optimized Mg-doped LMMP1 and LMMP3 are estimated to be 2–10 μm, respectively.

Figure 3b represents the energy dispersive spectroscopy (EDS) result of pristine LMP and depicts the compositional detail of the sample with elemental Mn(36.32% wt.), P(22.79% wt.), O(39.78% wt.) and impurity of Fe(0.01% wt.) is also detected. Similarly, Fig. 3d indicates the EDS result of 1% Mg-doped LMP, i.e. $LiMn_{0.99}Mg_{0.01}PO_4$ (LMMP1), which mainly consists of O(35.86% wt.), Mg(1.10% wt.), P(24.42% wt.) and Mn(39.68% wt.) and Fig. 3f depicts the EDS composition of $LiMn_{0.97}Mg_{0.03}PO_4$, which mainly consists of O(43.23% wt.), Mg(3.2% wt.), P(22.07% wt.) and Mn(33.94% wt.). Hence, EDS results reveal that the stoichiometry of all their compositions of pristine LMP, LMMP1 and LMMP2 remains almost intact in the final form of the phase after calcination.

3.3 Ionic Conductivity Measurement

Figure 4 depicts the measurement of the ionic conductivity of synthesized sample LMMP3 which is calculated with the help of conductivity and temperature. Measurement of activation energy is represented by the slope of this curve through which we can recurrently determine the kinetics of lithium-ion. Activation energy of the 3% Mg-doped LMP sample (LMMP3) was 0.8185 eV which clearly specifies the electronic conductivity (2.02×10^{-9}) of the sample. Further, the sample LMMP3 needs to be enhanced so that lithium-ion and electron kinetics become more boosted and also support in enhancement of electrochemical performances.

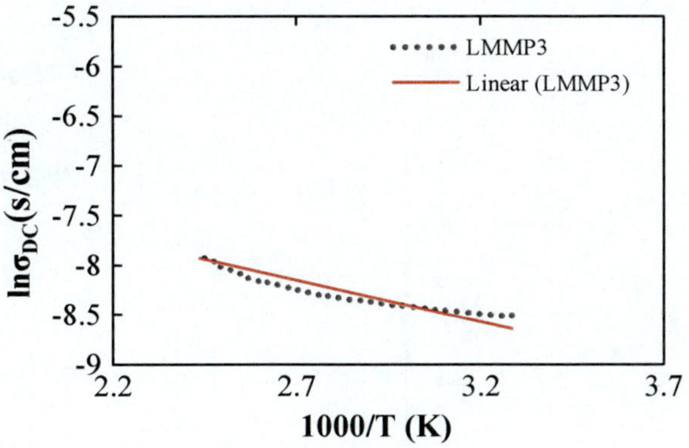

Fig. 4 Ionic conductivity of LMMP3 at a source voltage 1 V

3.4 Electrochemical Performance of Olivine ($LiMnPO_4$) Composites

Figure 5a shows the cyclic voltammetry curves at a scan rate of 0.05 mVs^{-1} versus 3% Mg-doped LMP composition Li^+/Li at room temperature. CV was performed within the potential range of 2.8–4.0 V. The CV curves also represent the formation of redox peaks at 2.9/2.7 V which can be allotted to Li-ion intercalation in the olivine structure. The first charge/discharge profiles of LMMP3 sample at 10 mA/g current density were displayed in Fig. 5b. The discharge capacities were observed to be 75 ± 5 $mAhg^{-1}$ for a sample LMMP3. The discharge plateau was obtained about 2.78 V during redox reactions as represented by the charge/discharge profiles of LMMP3. It might be due to variation in the oxidation state of Mn ion. These results for LMMP3 sample show good kinetics inside the cathode and are quite consistent

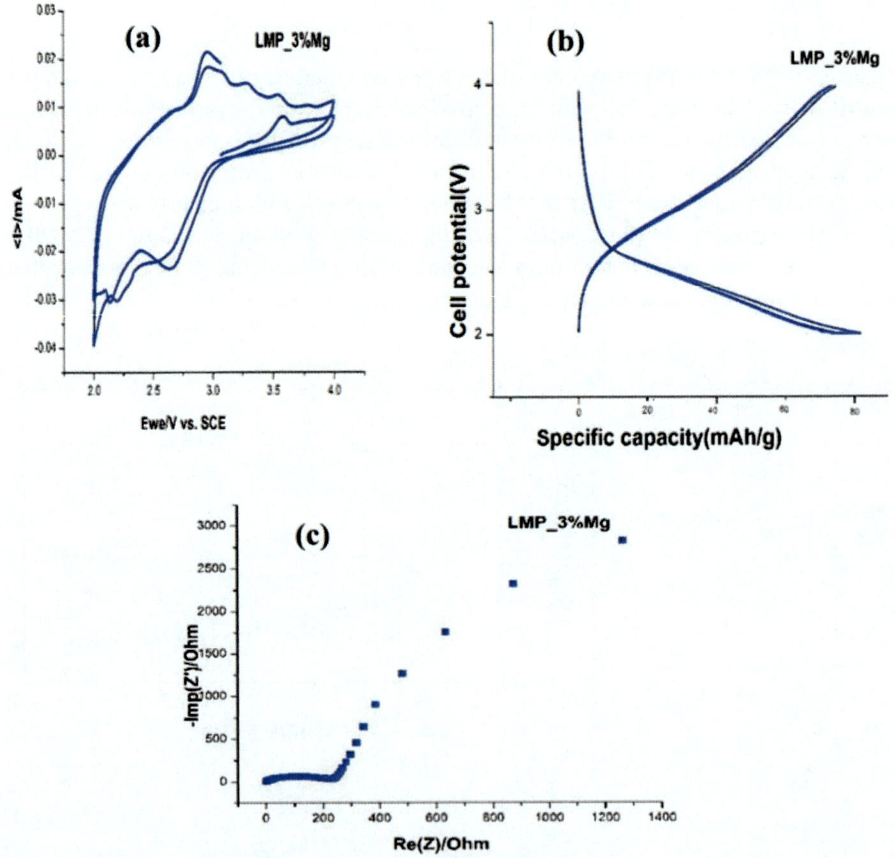

Fig. 5 **a** Cyclic voltammogram (2.8–4.0 V range), **b** charge/discharge curve and, **c** Nyquist plot from EIS spectra of LMMP3 sample

with the CV results. The changes in electrode impedance were analyzed at room temperature by performing Electrochemical impedance spectroscopy (EIS) for the synthesized composition of LMMP3. Figure 5c demonstrates the EIS curve between 10 mHz-100 kHz frequency range using an AC pulse with 5 mV amplitude. It can be analyzed from the curve that there is a semicircle observed in the high-frequency region followed by a straight line in the low-frequency region. The semicircle indicates the charge transfer resistance (R_{ct}) occurring in the electrode/electrolyte interface during the electrochemical reaction. The low-frequency region value inclined in the Warburg impedance (Z_w) represents the lithium-ion diffusion in the active cathode material. Hence, the lithium-ion diffusion coefficient (D) of the synthesized can be calculated with the help of standard equation and it is estimated, $3.58 * 10^{-14}$ cm^2/S.

4 Simulation Analysis

The simulation shows the distribution of lithium concentration by the end of 2700 s at the surface of electrode particles. The initial lithium concentration in the porous electrode is 1000 mol/m^3. Throughout the discharge process, as lithium is dissolved on the lithium metal electrode, the lithium concentration in the porous electrode increases.

Figure 6 reveals that the lithium-ion concentration at the surface of electrode particles is time-dependent in the battery during discharging. This helps to reveal that lithium-ion concentration is changing during the given time of 2700 s. Hence, red colour depicts that maximum lithium-ion reached towards the electrode and

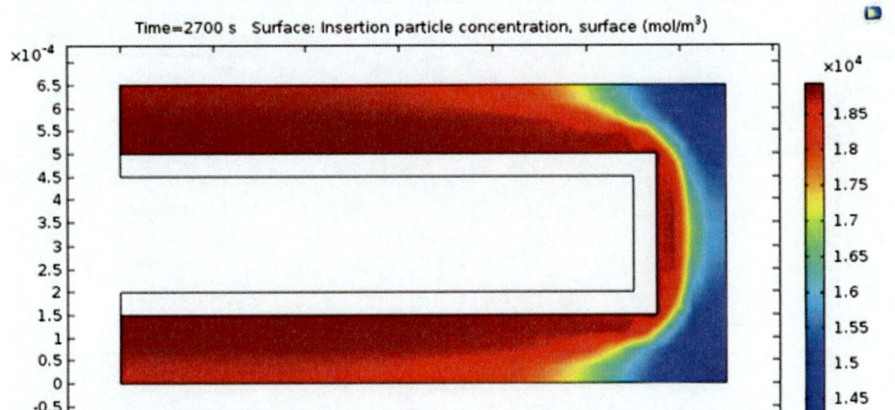

Fig. 6 The concentration of lithium at the surface of electrode particles

reaction initially happens at the boundary, then it starts expanding away from the boundary.

Figure 7 shows the cell voltage predicted by the model during the discharge simulation. In this, the maximum potential is found inside the inner portion of the electrolyte that depicts red colour and minimum potential is represented by the blue colour. In Fig. 8-time-dependent cell voltage has been analyzed in the simulated model during the discharge cycle of the battery. The x-axis represents electric potential on the

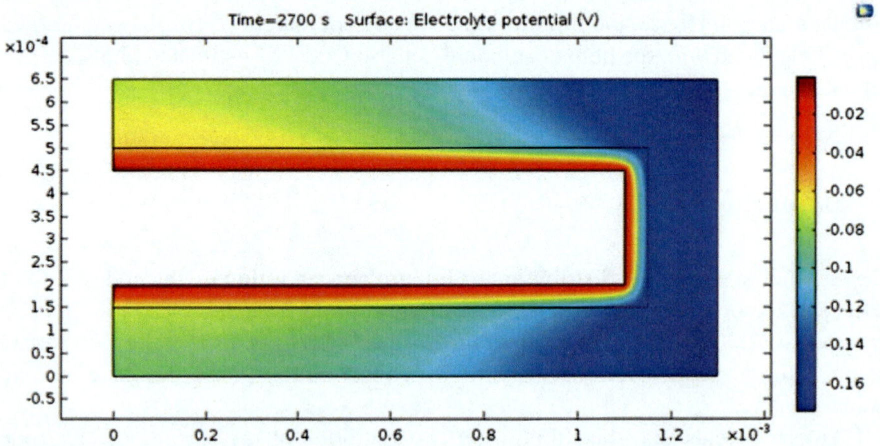

Fig. 7 The time-dependent cell voltage predicted by the simulated model during the discharge

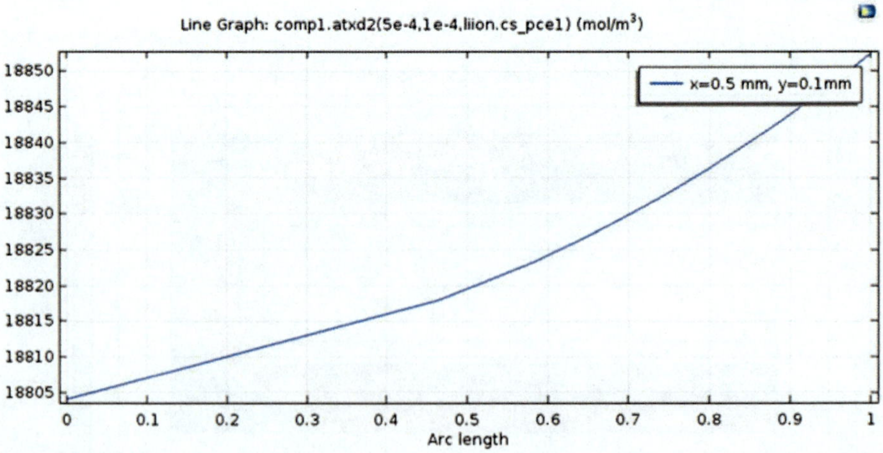

Fig. 8 Lithium concentration in the electrode particles at selected positions

boundary whereas y-axis represents the arc length. The blue line displays the concentration at the farthest located point away from the current collector and shows that minimum discharge has taken place at that point or the lower concentration.

This study provides a logical path for the design of components and construction of the structures of cathode materials such that significant improvement could be made for the lithium-ion batteries.

5 Conclusion

In this investigation, synthesis and characterization of pristine LMP and Mg-doped LMP at different wt% have been carried out using a solid-state route. XRD pattern showed the formation of a single phase of LMP and Mg-doped LMP samples. Non-uniformity in grain size of 2–10 μm was observed in synthesized samples. Activation energy and maximum discharge capacity of 3% Mg-doped samples are found approximately 0.818 eV and 75 ± 5 $mAhg^{-1}$ at 10 mA/g by the electrical and electrochemical performance of the sample LMMP3. The theoretical simulation study helps to provide logical and lucid direction for the efficient design of the structure of cathode material as an electrode and its composite cathode materials such that significant improvement could be made for implementation in Li-ion batteries.

References

1. Yang, G., Ni, H., Liu, H., Gao, P., Ji, H., Roy, S., Pinto, J., Jiang, X.: J. Power Sour. **196**(10), 4747–4755 (2011)
2. Wang, L., Zhang, H., Liu, Q., Wang, J., Ren, Y., Zhang, X., Yin, G., Wang, J., Zuo, P.: ACS Applied Energy Mater. **1**(11), 5928–5935 (2018)
3. Zhao, M., Fu, Y., Xu, N., Li, G., Wu, M., Gao, X.: J. Mater. Chem. A **2**(36), 15070–15077 (2014)
4. Zaghib, K., Guerfi, A., Hovington, P., Vijh, A., Trudeau, M., Mauger, A., Goodenough, J., Julien, C.: J. Power Sour. **232**, 357–369 (2013)
5. Lu, Q., Hutchings, G.S., Zhou, Y., Xin, H.L., Zheng, H., Jiao, F.: J. Mater. Chem. A **2**(18), 6368–6373 (2014)
6. Neef, C., Jähne, C., Meyer, H.-P., Klingeler, R.D.: Langmuir **29**(25), 8054–8060 (2013)
7. Ramar, V., Balaya, P.: Phys. Chem. Chem. Phys. **15**(40), 17240–17249 (2013)
8. Liu, S., Fang, H., Yang, B., Yao, Y., Ma, W., Dai, Y.: J. Power Sour. **230**, 267–270 (2013)
9. Kwon, N.H., Yin, H., Vavrova, T., Lim, J.H., Steiner, U., Grobéty, B., Fromm, K.M.: J. Power Sour. **342**, 231–240 (2017)
10. Köntje, M., Memm, M., Axmann, P., Wohlfahrt-Mehrens, M.: Prog. Solid State Chem. **42**(4), 106–117 (2014)
11. Chen, G., Shukla, A.K., Song, X., Richardson, T.J.: J. Mater. Chem. **21**(27), 10126–10133 (2011)
12. Cao, Y., Duan, J., Hu, G., Jiang, F., Peng, Z., Du, K., Guo, H.: Electrochim. Acta **98**, 183–189 (2013)
13. Di Lecce, D., Hu, T., Hassoun, J.: J. Alloy. Compd. **693**, 730–737 (2017)
14. Saroha, R., Gupta, A., Panwar, A.K.: J. Alloy. Compd. **696**, 580–589 (2017)

15. Saroha, R., Panwar, A.K., Sharma, Y.: Ceram. Int. **43**(7), 5734–5742 (2017)
16. Jain, A., Saroha, R., Pastor, M., Jha, A., Panwar, A.: Curr. Appl. Phys. **16**(8), 859–866 (2016)
17. Saroha, R., Panwar, A.K., Sharma, Y., Tyagi, P.K., Ghosh, S.: Appl. Surf. Sci. **394**, 25–36 (2017)
18. Saroha, R., Panwar, A.K.: J. Phys. D Appl. Phys. **50**(25), 255501 (2017)
19. Wu, W., Xiao, X., Huang, X.: Electrochim. Acta **83**, 227–240 (2012)
20. Golmon, S., Maute, K., Dunn, M.L.: J. Power Sour. **253**, 239–250 (2014)